Nuclear Weapons and the Environment

Environment and Society

Series Editor: Douglas Vakoch

As scholars examine the environmental challenges facing humanity, they increasingly recognize that solutions require a focus on the human causes and consequences of these threats, and not merely a focus on the scientific and technical issues. To meet this need, the Environment and Society series explores a broad range of topics in environmental studies from the perspec-tives of the social sciences and humanities. Books in this series help the reader understand contemporary environmental concerns, while offering concrete steps to address these problems.

Books in this series include both monographs and edited volumes that are grounded in the realities of ecological issues identified by the natural sci-ences. Our authors and contributors come from disciplines including but not limited to anthropology, architecture, area studies, communication studies, economics, ethics, gender studies, geography, history, law, pedagogy, phi-losophy, political science, psychology, religious studies, sociology, and theology. To foster a constructive dialogue between these researchers and environmental scientists, the Environment and Society series publishes work that is relevant to those engaged in environmental studies, while also being of interest to scholars from the author's primary discipline.

Titles in the series:

Living Deep Ecology: A Bioregional Journey, by Bill Devall and edited and introduction by Sing C. Chew
Environmental and Animal Abuse Denial: Averting Our Gaze, edited by Tomaž Grušovnik, Reingard Spannring, and Karen Lykke Syse
Conservation, Sustainability, and Environmental Justice in India, edited by Alok Gupta
Secular Discourse on Sin in the Anthropocene: What's Wrong with the World? by Ernst M. Conradie
Environment, Social Justice, and the Media in the Age of the nthropocene, edited by Elizabeth G. Dobbins, Maria Lucia Piga, and Luigi Manca
Global Perspectives on Eco-Aesthetics and Eco-Ethics: A Green Critique, edited by Krishanu Maiti and Soumyadeep Chakraborty
Australian Wetland Cultures: Swamp Thinking, edited by John Charles Ryan and Li Chen
Motor Vehicles, the Environment, and the Human Condition: Driving to Extinction, by Hans A. Baer
Ecology, Artificial Intelligence, and Virtual Reality: Life in the Digital Dark Ages, by Sing C. Chew
Portland's Good Life: Sustainability and Hope in an American City, by R. Bruce Stephenson
Nuclear Weapons and the Environment: An Ecological Case for Non-Proliferation, by John Perry

Nuclear Weapons and the Environment

An Ecological Case for Non-Proliferation

John Perry

LEXINGTON BOOKS
Lanham • Boulder • New York • London

Published by Lexington Books
An imprint of The Rowman & Littlefield Publishing Group, Inc.
4501 Forbes Boulevard, Suite 200, Lanham, Maryland 20706
www.rowman.com

6 Tinworth Street, London SE11 5AL, United Kingdom

British Library Cataloguing in Publication Information Available

Library of Congress Cataloging-in-Publication Data

Name: Perry, John, 1946–, author.
Title: Nuclear weapons and the environment : an ecological case for non-proliferation / John Perry.
Description: Lanham, Maryland : Lexington Books, 2021. | Series: Environment and society | Includes bibliographical references and index. | Summary: "In Nuclear Weapons and the Environment, John Perry highlights the grave risks associated with the continued proliferation of nuclear weapons, pointing not just to the dangers to human life but to the severe environmental damage caused by nuclear device testing"—Provided by publisher.
Identifiers: LCCN 2020055097 (print) | LCCN 2020055098 (ebook) | ISBN 9781793602831 (cloth) | ISBN 9781793602848 (epub)
Subjects: LCSH: Nuclear weapons—Environmental aspects. | Nuclear weapons—Testing—Environmental aspects. | Nuclear energy—Environmental aspects. | Nuclear nonproliferation.
Classification: LCC U264 .P46 2021 (print) | LCC U264 (ebook) | DDC 363.738—dc23
LC record available at https://lccn.loc.gov/2020055097
LC ebook record available at https://lccn.loc.gov/2020055098

Contents

Preface

The Ultimate "What If": Nuclear Deterrence and the Environment

Recent environmental challenges have captured the world's attention: global warming, the COVID-19 pandemic, and our dependence on fossil fuels, to mention a few. But the very real possibility of nuclear war remains an elusive but very real threat to all of us. While the risk of an "all-out" conflagration between the super powers may have diminished with the end of the Soviet Union, the possibility of a regional nuclear war is a real possibility. A nuclear war between India and Pakistan, for example, would have global and catastrophic consequences. As well, terrorists seek nuclear devices such as 'dirty bombs' in order to achieve their objectives.

The thesis of this book is that the proliferation of nuclear weapons is a grave risk to the human community and its environment. The unstable political situation between not only India and Pakistan but also North Korea and South Korea/the United States of America, among other troubled relationships now involve not only conventional weapons but also nuclear ones as well due to nuclear proliferation.

Other issues to do with nuclear proliferation include the possible theft of nuclear devices by non-state actors such as Al-Qaeda and the Taliban, the remediation of nuclear test sites, and the risk of accidents happening in places where these weapons are produced. Recent disasters at two nuclear power facilities at Chernobyl in the Ukraine and Fukushima Daiichi in Japan have underlined new levels of safety needed in the power industry if it and we are to survive.

Now in its fiftieth year the Nuclear Non-Proliferation Treaty and its additional protocols have served the human community well, but now faces a number of challenges going into the future.

While specialized and technical treatments of all these topics are needed, the present volume seeks to offer a wider readership with an introduction to these topics and dangers with the hope that nuclear safety and the importance of non-proliferation would remain a central concern for us all.

ONE

Introduction

Nuclear Weapons and the Environment: An Ecological Case for Non-Proliferation

On July 7, 2017 following a decade of advocacy by the International Campaign to Abolish Nuclear Weapons and the International Physicians for the Prevention of Nuclear Weapons and other like-minded groups most of the world's nations agreed to embrace a landmark global agreement to ban nuclear weapons. This agreement is the Treaty on the Prohibition of Nuclear Weapons approved by 122 countries in 2017 and now signed by seventy nations. The treaty emerged from the Humanitarian Initiative, a group of non-nuclear states, which highlights the horrendous consequences of nuclear war. They took the decision to shift the argument over nuclear weapons from the Security Council to the UN General Assembly, which votes by majority. The treaty will probably enter into force once fifty states have ratified it.[1]

Nuclear weapons have the potential to devastate the world with their persistent environmental effects and thereby to create widespread humanitarian destruction. Alan Robock, the co-author of a study on the serious environmental effects of a limited nuclear war between Pakistan and India published in the *Journal of Geophysical Research*, told the *Guardian* newspaper that "Nuclear weapons are the greatest environmental danger to the planet from humans—not global warming or ozone depletion."[2] Until the treaty these weapons were not subject to a comprehensive ban, and this was a glaring shortcoming in international law.

The Treaty on the Prohibition of Nuclear Weapons bans nations from developing, testing, producing, manufacturing, transferring, possessing, stockpiling, using or threatening to use nuclear weapons, or allowing nuclear weapons to be placed on their territory.[3] It also prohibits them from assisting, encouraging or inducing anyone to engage in these activities.

Both the Pentagon and the State Department in the United States stand opposed to the Treaty and the UN's humanitarian campaign because it seemed to them to be an effort to delegitimize nuclear deterrence on which US security and alliances depended.[4]

The treaty represents a global effort to halt nuclear proliferation. Nuclear weapons are the only devices known to man/woman with the potential of destroying all complex forms of life on Earth within a relatively short period of time. A war fought using one hundred nuclear weapons which represent a small percentage of the total global stockpile of 13,500 nuclear warheads—would make our planet uninhabitable.[5]

A regional nuclear war would disrupt the global climate and agricultural production. This might mean that two billion people would be at risk of famine. The smoke and dust from a limited nuclear war would cause an abrupt drop in global temperatures and rainfall by blocking up to 10 percent of sunlight from the Earth's surface.

A nightmare preview of the devastation that can be caused by massive fires happened in the last year of the Second World War. The final crisis for Japan took place earlier than the nuclear attacks on Hiroshima and Nagasaki. It began on the evening of March 9–10, 1945, when Maj. Gen. Curtis Lemay launched a massive, low-level incendiary attack on Tokyo conducted by three hundred B-29 bombers. These bombers dropped 8,500 bombs that, in turn, dispersed a half-million canisters of incendiary fuel. By dawn, sixteen square miles of Tokyo were completely devastated, eighty-eight thousand people were dead, and the survivors were burned and disfigured.[6]

But in the subsequent study of this event as well as what happened at Hiroshima and Nagasaki later in 1945 the terrible environmental effects of fire were not considered seriously. Immediately following the end of World War II, the US government greatly underestimated the damage that nuclear weapons would cause. In devising its plans to fight strategic war after World War II the US government developed detailed knowledge about the blast damage caused by nuclear weapons but failed to develop knowledge about an even more devastating effect, the mass fire, popularly known as a "firestorm," such as the fire that burned down Hiroshima in August, 1945.[7] Sadly, what happened in Japan in 1945 pales into insignificance before the prospect of a nuclear war. The smoke and dust from a regional nuclear war would create what is popularly known as a "nuclear winter."

Only in the early 1980s did scientists begin to realize what a nuclear war would do to our planet as a whole. This began in 1971 when a small group of planetologists who had gathered to analyze the results of the *Mariner 9* observations on Mars found that the entire planet was covered by a huge dust storm that lasted three months. They calculated how such a long-lasting dust cloud would change the conditions on the Martian surface. The conclusion was reached that it would lower the ground temperature to a significant degree.[8] Having examined the meteorological records on earth they began to understand that every time a major volcano had erupted on planet earth over the past several centuries there has been a small drop in the global temperature, lasting a year or more.

The original, informal group of scientists who had been involved in the Mariner project, who called themselves TTAPS after the first letters of their last names, were later shown a paper written in 1982 by two scientists working in the Max Planck Institute in West Germany. This study calculated that the massive forest fires triggered by nuclear blasts would inject several hundred million tons of smoke into the atmosphere in a nuclear war and that the smoke "would strongly restrict the penetration of sunlight to the earth's surface."[9]

In 1983, the TTAPS group estimated the effect that the smoke from burning cities and the dust from ground bursts would have on our environment. They concluded that a major nuclear exchange would cover at least the northern hemisphere, and perhaps the entire planet, with a pall of smoke and dust. This would plunge the surface into virtual darkness for up to six months and cause the temperature to drop by up to 40 degrees centigrade.[10]

Sudden global cooling would shorten growing seasons, threatening agriculture worldwide. Increases in food prices would create serious issues for hundreds of millions of the poorest people of the world who would find food too expensive for them to purchase. For those already suffering from malnourishment a 10 percent decline in food consumption would lead to starvation. They would also have to face infectious disease pandemics and conflict over scarce resources.[11] New research done by The International Physicians for the Prevention of Nuclear War suggests strongly that the climate effects of the use of nuclear weapons in a nuclear war using as few as one hundred weapons anywhere in the world would disrupt the global climate and agricultural production so severely that the lives of more than two billion people would be in danger.[12]

Corn production in the United States would decline by an average of 10 percent for an entire decade, with the most severe decline (20 percent) in year five after the war. Soybean production would also decline by about 7 percent with the most severe loss, more than 20 percent, in year five.

China is of special concern. There would be a significant decline in middle season rice production in China. Chinese winter wheat production would fall 50 percent in the first year after the war. More than a billion people in China alone would face severe food insecurity.

The nuclear winter is well described by Dr. Ira Helfand, the author of the report on the effect on food production by a regional nuclear war. He summarizes the situation in China in the following way: "The prospect of a decade of widespread hunger and intense social and economic crisis in the world's largest country has immense implications for the entire global community, as does the possibility that the declines in Chinese wheat production will be matched by similar declines in other wheat producing countries."[13]

The dire consequences of a nuclear war and the subsequent nuclear winter were confirmed by the Soviet Union in the midst of the Cold War by using the BESM-6 computer at the Soviet Academy of Science Computer Center in Moscow. Nuclear strikes in the Northern hemisphere would, this computer calculated, cause climate change in the Southern hemisphere. Soviet calculations predicted that the spring and summer in Western Europe would turn into a severe winter, and the winter into an arctic winter; the temperatures would drop by 15–40 degrees Celsius over the region extending from Chad to Novosibirsk, the Caspian Sea to Sri Lanka, including Pakistan and Western China.

BESM-6 revealed that a completely different planet would come into being after a nuclear war. The computer found that conifer tree forests would be worst affected by the radiation, and that birds would be impacted worse than insects. Radioactivity would be carried not only by rain clouds but also by pollen that could be inhaled by mammals. The landscape after the nuclear attack would be one with few conifers, but many shrubs, few mammals, but a lot of insects.[14]

The potential harm to the world's environment of a nuclear war is both morbid and consequential. It is one thing to know that nuclear weapons are enormously destructive to the environment. It is another thing to have a detailed picture of the damage that would result from their use.

The current political situation between India and Pakistan remains unstable, and an open war almost occurred in February 2019. Both countries have nuclear weapons, India 130–140 warheads and Pakistan 140–150 warheads. What would happen to planet Earth if they went to war again and made use of the weapons of mass destruction that they possess?

Chapter 2 will present historical background including the use of nuclear weapons on Hiroshima and Nagasaki in 1945. The efforts of Niels Bohr, Eugene Wigner, James Franck, and Leo Szilard to prevent proliferation of the atom bomb to the Soviet Union will be discussed.

Chapter 3 discusses the history, success and limitations of the Nuclear Non-Proliferation Treaty which celebrated its fiftieth anniversary on March 5, 2020. Nine countries at present have approximately 15,000 nuclear weapons. Five European nations host US nuclear weapons and twenty-five others, including Canada, endorse the use of nuclear weapons. Are the treaty and the current legal safeguards now in force adequate? Chapter 4 has to do with nuclear proliferation in India and Pakistan and the risk that a nuclear war could break out between these two traditional enemies. The role of A. Q. Khan in assisting Pakistan to develop the bomb will also be discussed. Chapter 5 will delve into some of the environmental and human costs of producing nuclear weapons. There have been several particularly devastating accidents in the United States and in what is now Russia in factories producing plutonium for nuclear warheads, and these will be examined. Two recent environmental disasters due to nuclear accidents in order of seriousness are the following: Chernobyl and Fukushima Daiichi. The first will be dealt with in chapter 6 and the second in chapter 7. Chapter 8 studies the apparently successful efforts at non-proliferation after 1994 following the dissolution of the Soviet Union. Then US Secretary of Defense William J. Perry negotiated a complex series of agreements to do this. This project moved all the nuclear warheads stored in the Ukraine, Kazakhstan, and Belarus to Russia for "safe-keeping."

Chapter 9 will deal with the difficult problem of the potential risk of proliferation of weapons of mass destruction by non-state actors through theft, terrorism, and other means such as neglect. The important topic of nuclear weapons testing, which means inflicting the conditions of war on fragile environments, will be explored in chapter 10. Chapter 11 will discuss Pakistan as the biggest exporter of nuclear technology in the history of the nuclear age. And chapter 12 will discuss one of the nations to which Pakistan's exported nuclear technology, namely North Korea, and it will carry the title of the Ultimate Weapon as the Ultimate Guarantee of Regime Survival: North Korea and the Bomb.

Ever since the use of the atomic bombs at Hiroshima and Nagasaki, nuclear politics have presented various challenges to our environment if it is to survive. If we are to make the correct choices now, we need to understand how the present nuclear proliferation has occurred and whether the new Treaty on the Prohibition of Nuclear Weapons will be able to contain it and to minimize the risks we all now face.

NOTES

1. *The Economist. The World in 2020*, p. 72.
2. *The Guardian,* Dec. 12, 2006, http://www.theguardian.com/environment/2006/dec/12/nuclearindustry.climatechange, visited Dec. 2019.
3. United Nations Office for Disarmament Affairs https://www.un.org/disarmament/wmd/nuclear/tpmw.
4. Tanenwald, Nina.2018. "The Vanishing Nuclear Taboo? How Disarmament Fell Apart," *Foreign Affairs*, V. 97, Number 6, November/December, p. 21.
5. www.icanw.org/the-facts/climate-disruption-and-nuclear-famine,
6. Reed, Thomas C. and Stillman, Danny B. 2009. *The Nuclear Express. A Political History of the Bomb and Its Proliferation.* Minneapolis: Zenith Press, p. 20.
7. Eden, Lynne. 2004. *Whole World on Fire. Organizations, Knowledge, and Nuclear Weapons Devastation.* Ithaca and London: Cornell University Press, p. 1.
8. Dyer, Gwynne. 2016. *War. The New Edition.* Toronto: Random House, p. 340.

9. Ibid, p. 341.
10. No Taps for TTAPS (nuclear-winter hypothesis), 1985. *Scientific American,* Feb. 1985, Vol. 252, pp.61–62.
11. Ibid.
12. https://www.ippnw.org/nuclear-famine.html.
13. Dr. Ira Helfand, https://www.psr.org./wp-content/uploads/2018/04/two-billion-at-risk.
14. Rindzevičuté. Eyle, 2016, *The Power of Systems: How the Policy Sciences Opened up the Cold War,* 2016, Ebook.

TWO

Historical Background

The literal meaning of Hiroshima is "broad island."[1] This meaning, of the city, however, does not convey its importance for us today because Hiroshima evokes for us the nuclear nightmare.

Many have pointed out that the statistics concerning the power of the blast, or the number of people killed, miss the point of the name Hiroshima. It carries its own urgency and poignancy that other cities which underwent destruction during the Second World War such as Dresden, Hamburg or Tokyo do not. The maximum number of deaths for an individual city during the Second World War was about 200,000, and the main victims were Dresden, Hiroshima and Nagasaki, the principal targets of Allied terror bombing.[2]

Along with her sister city Nagasaki, conveyed in the name Hiroshima is a sense of totality and of ultimacy of annihilation and destruction not only of these two cities but potentially of the whole world, of all nations and cities. The name Hiroshima and the encounter of its people with the atomic bomb have immediate importance for us all in the nuclear age in which we now live. In a specific way Hiroshima and Nagasaki are cities of the dead, not in a metaphorical sense but in reality.

People used the Japanese word *bukimi* meaning weird to describe, on the one hand, the continuing good fortune of Hiroshima because before August 6 it had not yet suffered aerial bombardment and, on the other hand, the expectation of the tragic fate that lay ahead.[3]

Leaflets had been dropped on July 27 threatening Hiroshima with total destruction if Japan did not surrender immediately. No mention was made of an atomic bomb or any other weapon. The people of Hiroshima did not receive any warning about the atomic bomb because American policy makers for various strategic reasons chose not to give them any prior warning.

Another reason for surprise was that after two air raid alerts had sounded during the night of August 5 the sirens had sounded a third time at 7:10 am on August 6 because of planes sighted over southern Japan. Sometime later a single B-29 approached the city, which was in fact the atomic bomb mission's weather plane, but since it quickly departed the all-clear was sounded at 7:32 am. Shortly after 8 am, air defense spotters saw two additional B-29s, which were the planes which carried out the atomic bombing, heading for Hiroshima but no additional air raid warning was sounded. A radio broadcast spoke about the planes urging that the people should take shelter should the planes appear over the city but then added the reassuring idea that that they seemed only on a reconnaissance mission.[4]

People were unprepared for the atomic bomb for several reasons; the relaxation induced by the all clear signal, the sense that they were somehow especially protected, and most importantly the incapacity to conceive of the unprecedented nightmare of the weapon that was about to strike them.

Only those at some distance from the explosion could clearly make out the sequence of the great flash of light accompanied by the terrific heat of the fireball, then the sound and force of the blast, and finally the striking multicolored cloud rising above the city.

A history professor who witnessed the explosion from a suburb three miles away wrote about the awful beauty of the spectacle:[5]

> A blinding . . . flash cut sharply across the sky. . . . I threw myself on the ground . . . in a reflex movement. At the same moment as the flash, the skin of my body felt a burning heat. . . . [Then there was] a blank in time . . . dead silence . . . probably a few seconds . . . and then a huge "boom" . . . like the rumbling of distant thunder. At the same time a violent rush of air pressed down my entire body. . . . Again there were some moments of blankness . . . then a complicated series of shattering noises. . . . I raised my head, facing the center of Hiroshima to the west. . . . [There I saw] an enormous mass of clouds . . . [which] spread and climbed rapidly . . . into the sky. Then its summit broke open and hung over horizontally. It took the shape of . . . a monstrous mushroom with the lower part as its stem-it would be more accurate to call it the tail of a tornado. Beneath it more and more boiling clouds erupted and unfolded sideways. . . . The shape . . . the color . . . the light . . . were continuously shifting and changing.

The bomb was completely on target and when it exploded carried a force of 20,000 tons of TNT, eighteen hundred feet in the air close to the center of a city built mainly of wood. Its area of destruction was more than two miles in all directions and destroyed sixty-thousand of the ninety-thousand buildings within more than three miles, an area close to the extent of the city's limits. Flash burns from the heat generated by the release of a large amount of radiant energy occurred at distances of more than two and a half miles. Injuries from the blast and from splintered glass and falling debris, occurred throughout the city and beyond.

The number of deaths, immediately and over the time that followed, will never be known. An accurate number may never be calculated but what can be said is that all of Hiroshima immediately became involved in the atomic disaster.

1.2 miles is considered to be the crucial radius for suffering radiation effects, and for a high mortality rate from blast, heat, or radiation—though many were killed outside this radius. Within it, at points near its hypocenter, heat was so extreme that metal and stone melted, and human beings were literally incinerated. The area was enveloped by fires fanned by a violent "firewind"; these broke out almost immediately within a radius of up to two miles.[6]

Robert Jay Lifton stated that people could recall a sudden flash, an intense sensation of heat and finding themselves pinned under debris or simply awakening from a state of unconsciousness. "The most striking psychological feature of this immediate experience," he wrote, "was the sensation of a sudden and absolute shift from normal existence to an overwhelming encounter with death." This is well described by a shopkeeper's assistant, who was thirteen years old at the time of the bomb and fourteen hundred meters from the hypocenter.

> I was a little ill . . . so I stayed at home that day. . . . There had been an air-raid warning and then an all-clear. I felt relieved and lay down on the bed with my younger brother. . . . Then it happened. It came very suddenly. . . . It felt like an electric short-a bluish sparkling light. . . . There was a noise, and I felt a great heat-even inside of the house. When I came to, I was underneath the destroyed house. . . . I didn't know anything about the atomic bomb so I thought that some bomb had fallen directly upon me. . . . And then when I felt that our home had been directly hit, I became furious. . . . There were roof tiles and walls-everything black-entirely covering me. So I screamed for help . . . and from all around I heard moans and screaming, and then I felt a kind of danger to myself. . . . I thought that I too was going to die in that way. I felt this way at that moment because I was absolutely unable to do anything at all by my own power. . . . I didn't know where I was or what I was under. . . . I couldn't hear voices of my family. I didn't know how I could be rescued. I felt I was going to suffocate and then die, without knowing exactly what had happened to me. That was the kind of expectation I had.

When the bomb fell on Hiroshima Bernard Brodie was a young scholar who had just joined the Institute of International Studies at Yale University. It was Brodie who formulated a key concept based on what had happened at Hiroshima: "It is already known to us all that a war with atomic bombs would be immeasurably more destructive and horrible than any the world has known. That fact is indeed portentious, and to many it is overwhelming" he wrote, and that has changed the nature of war and the conditions of peace beyond recognition.[7]

The world was at war when the atomic bombs were developed, tested and used on Hiroshima and Nagasaki in August 1945. In the course of the Second World War, every major industrial nation began a program to build atomic bombs: the Germans, the British, the French before their surrender, the Soviets, the Americans, and the Japanese. But nuclear weapons development required a massive commitment of government funds, funds that would have to be diverted from their normal use in the war. Many realized that if atomic bombs could be built, they would be decisive, in which case no country engaged in combat could afford not to pursue them.[8]

The German program fragmented and stalled. The French program was stillborn. The Soviets, fighting for their lives against an almost overwhelming German invasion, put their early work on hold, only to revive it again in 1943 after their espionage revealed the extent of the developing programs in Britain and the United States. The Japanese realized that a bomb program was beyond their resources.

Three physicists, Albert Einstein, Eugene Wigner, and Leo Szilard, participated in a meeting that resulted in the Einstein-Szilard letter which prompted President Franklin D. Roosevelt to initiate what became known as the Manhattan Project to develop a nuclear bomb. President Roosevelt duly authorized a full-scale Anglo-American weapons program on October 9, 1941.

What follows is an effort to explain some of the efforts of three atomic scientists who worked on the bombs that destroyed Hiroshima and Nagasaki to deal with the possibility of the proliferation of the atom bomb to the Soviet Union. Leo Szilard, Niels Bohr, and Eugene Wigner attempted to prevent the use of the bombs against Japan and to place the fateful knowledge of how to construct them into the hands and minds of the international community.

Leo Szilard (1898–1964) would agree with Brodie's view of the capital importance of the atomic bomb but this was even before the Hiroshima event, not after it, as in the case of Brodie. Szilard was a Hungarian Jewish émigré who had just barely escaped the Third Reich and had later become the first to conceive of the idea of a nuclear chain reaction[9] while crossing a London street on a grey Depression morning in 1933. Szilard was frightened by the Nazis. In 1933 he fled from Berlin by train just twenty four hours before Julius Streicher, a key collaborator with Hitler, directed a national boycott of Jewish businesses with the effect that Jews were attacked in the streets, and thrown off trains while trying to leave Germany. After that event, Szilard always kept two suitcases packed and ready to take with him if it was necessary for him to flee again.[10]

Despite the strong dislike of General Leslie Groves, Szilard made significant contributions to the Manhattan Project, for which Groves was appointed as head and which developed the atom bomb.[11] Perhaps because he was a maverick Groves suspected Szilard of disloyalty to the project.12 As the final preparations were made to test the bombs Szilard argued that the bombs should be brought into being as quickly as possible because unless this were done and until they were used "the public would not comprehend their destructive power" and only such understanding would lead to a stable peace.[12]

Notwithstanding these earlier views the atomic bombing of Hiroshima and Nagasaki horrified Szilard. He felt guilty for the part he played in the development of such terrible weapons. Before the bombs fell on Hiroshima and Nagasaki, in July 1945 he had circulated a petition to the President of the United States among his fellow atomic scientists. Szilard had argued that large moral responsibilities devolved on the United States in consequence of its possession of the bomb:[13]

> The development of atomic power will provide the nations with new means of destruction. The atomic bombs at our disposal represent only the first steps in this direction, and there is almost no limit to the destructive power which will become available in the course of their future development. Thus a nation which sets the precedent of using these newly liberated forces of nature for purposes of destruction may have to bear the responsibility of opening the door to a new era of devastation on an unimaginable scale.

Szilard's visionary views on the moral responsibility of creating weapons of mass destruction and his quest for peace had an influence much later on a younger Russian scientist Andrei Sakharov who attributed the beginning of his courageous work of protest to reading a short story written by Szilard entitled "My Trial as a War Criminal," a chapter in his book *The Voice of Dolphins*, published in 1961 in the United States. It was about a fictionalized war between the USSR and the US, a very devastating one, which brought victory to the USSR. Szilard and a number of other physicists are put under arrest and then face the court as war criminals for having created weapons of mass destruction. Neither they nor their lawyers could create cogent proof of their innocence. Sakharov and others were amazed by the paradox. They could not get away from the fact that they were developing weapons of mass destruction. They thought it was necessary. Such was their inner conviction. But it did not allow Sakharov and others to live in peace with themselves.[14]

The second Los Alamos physicist who attempted to forestall an arms race for control of the bomb between the United States, Great Britain and Russia was Niels Bohr (1885–1962). Bohr was born in Denmark to a Jewish mother. In 1922, he won a Nobel Prize for explaining the structure of the atom. He was in Denmark at the time of the Nazi invasion of Western Europe in 1940, staying in Copenhagen until he learned in 1943 that he was about to be arrested. Bohr and his family escaped to Sweden in a fishing boat. He and his son were then flown to England in the bomb bay of a British Mosquito aircraft, crossing over the German controlled airspace over Norway along the way. In due course Bohr reached Los Alamos where his presence was so important that he was given a code name "Nicholas Baker."[15] Along with Albert Einstein Bohr was one of the two most famous physicists in the world at the time. In terms of the atomic bomb, Bohr was even more important than Einstein because Bohr played a central role in the understanding of nuclear fission, the way the nucleus of an atom may be split releasing vast amounts of explosive energy.[16] Bohr was generous, avuncular and completely devoid of those instincts for rivalry that can easily sour professional relationships.[17]

When Bohr reached Los Alamos he was persuaded that the peril of atomic war was common to all nations.[18] To the question of what defense was necessary to remove the risk of using an atomic weapon Bohr replied: "only international co-operation, exchange of scientific discoveries, internationalization of the achievements of science can lead to the elimination of the very necessity to use the atomic bomb. This is the only rightful method of defense. . . . All scientists believe that this greatest discovery must belong to all nations and serve the unprecedented progress of humankind. Atomic energy, once discovered, cannot remain the asset of one nation since a country which does not possess this secret can soon discover it independently."[19]

Bohr's vision of an open world pioneered by scientific sharing seemed to be an appeal for collaboration. Shortly after his arrival in the US Bohr renewed his acquaintance established in Oxford with Felix Frankfurter, now a justice of the Supreme Court. Frankfurter was a good friend of US president Franklin Delano Roosevelt.

The first encounter between Bohr and Frankfurter took place over lunch in the justice's office in the Supreme Court building in Washington. Referring to his work on the atomic bomb in Los Alamos as Project X Bohr told Frankfurter that many physicists, including himself, were disturbed by what they were doing and that they had asked him to speak on their behalf with those in power. Bohr hoped that Frankfurter would approach Roosevelt about what was troubling so many involved in Project X.

Bohr speculated that the US had rushed into building the bomb without considering the long term implications of nuclear power.[20] As a group the scientists were troubled by the fact that the more progress they made on Project X and the closer they came to success, the more concerned they were about these long term consequences.[21] Bohr argued that a new international order was necessary. Atomic weapons were so powerful that they could only be controlled in an "open world." Only if such openness were achieved could we be confident that any real or potential enemy nations were not engaged in stockpiling nuclear weapons. Scientists had developed an intellectual culture which allowed them to check on each other's results, and this was how scientific progress such as had happened with fission could be achieved. Bohr and other physicists were concerned about the threat of nuclear annihilation.[22] Bohr argued that such an international agreement could only be attained if the Soviets were brought into the discussion before the bomb's development was completed and before hostilities were over.[23] In his presentation to Frankfurter Bohr assumed two things. First, the bomb was beyond anything previously experienced by humans. And second, it could not be monopolized. Rival nations would sooner or later obtain it. Bohr understood that the bomb would dramatically alter the way politics operated; there would only be co-operation or disaster.

Bohr's analysis was that Stalin must be informed about the Manhattan Project *before* its development had gone so far as to make any discussion seem coercive rather than friendly. It was important that the Russians understood that an Anglo-American atomic monopoly was not being set up against them. His proposal was simply that the Russians should be informed that the Manhattan Project existed, but only this. If the Soviet response to this information were positive, then further co-operation could be considered. But further information would be withheld until the safety issues were addressed.[24]

Bohr's final point was important. He held the view that the international community of scientists could play an important role beyond merely a technical one. "Helpful support," he told Frankfurter, "may perhaps be afforded by the worldwide scientific collaboration that had embodied such bright promises for common human striving. On this background personal connections between scientists of different nations might even offer a means of establishing preliminary and non-committal contact." Bohr imagined the possibility of re-creating the collaboration between scientists of all nations that had existed in his own institute in Copenhagen before the war.[25]

\In the last week of March, 1944 Frankfurter was able to have an hour and a half meeting with President Roosevelt at which he gave his friend a detailed account of Bohr's analysis focusing on the central worry: "it would be disastrous if Russia should learn on her own about X rather than that the existence of X should be utilized by this country and Great Britain as a means of exploring the possibility of an effective international arrangement with Russia for dealing with the problems raised by X."[26] Frankfurter also argued with the president that Bohr was certain that Soviet physicists would sooner or later arrive at their own information necessary to build their own atomic weapons.

Bohr had identified the central issue as coercion. Of course, there were a number of other reasons which would enhance the advantages of telling the Russians in good time, with the possibility that the disclosure might bring with it good will. If the bomb were kept secret, and this gave the Allies an advantage, how would that advantage operate? Would such coercion work to intimidate the Soviets? At least theoretically, an approach to Stalin offered benefits. A proposal for the international control of atomic energy could form the beginning of great power collaboration after the war.

Following Frankfurter's long conversation with Roosevelt about Bohr's ideas the president asked Frankfurter to request Bohr on his next visit to Britain to "inform our friends in London that the President was most eager to explore the safeguards in relation to X."[27] Bohr prepared a document in preparation for a meeting with the British Prime Minister Churchill which was intended for Sir John Anderson, the British cabinet minister responsible for science. In it he suggested that they look on the Manhattan Project as a global responsibility rather than as a threat. Bohr concluded the document by saying: "Such an initiative, aimed at forestalling a fateful competition about the formidable weapon, need in no way impede the importance of the project for the immediate military objectives, but should serve to uproot any cause for distrust between the powers on whose harmonious collaboration the fate of the coming generations will depend."[28]

The meeting between Bohr and Churchill on May 16, 1944 did not go well. Churchill cut Bohr short before the Nobel Laureate could even get to his main point. "I cannot see what you are worrying about," Churchill grumbled. The atom bomb was only a bigger version of existing bombs and "made no difference to the principles of war." The Prime Minister went on to say that there were no long-term problems with the bomb that could not be amicably settled between myself and my friend President Roosevelt."[29]

Bohr finally had his meeting with Roosevelt later on August 26, 1944. Bohr found the President "very agreeable" and agreed with him that contact with the Soviet Union "had to be tried along the lines he suggested." The president went so far as to suggest that he was optimistic that such an approach would have a "good result" and open "a new era."[30] Roosevelt added that in his opinion Stalin was enough of a realist to grasp the revolutionary importance of the bomb and its consequences. No less important, Roosevelt told Bohr, was the fact that he was confident that he would win the prime minister to this view. "They had disagreed before," he said, "but in the end had always succeeded in resolving their differences." He suggested he and Bohr might meet again after the upcoming Quebec Conference but, in the meantime, if Bohr had any further views to communicate, "he would welcome a letter."[31]

How realistic was Bohr being? On June 13, 1944, weeks before Bohr met the president, Roosevelt and Churchill had signed an Agreement and Declaration of Trust, which specified that Britain and the United States would work together in seeking to control available supplies of uranium and thorium ore during and after the war. The Agreement was designed to ensure that, if an arms race did develop after the war, the United States would always have enough supply of raw material "to ensure superiority."[32]

At a meeting after the Quebec Conference on September 19, 1944 Churchill won over the president convincing him of the need for secrecy, arguing that Bohr's attitude, though shared by many other scientists and many officials surrounding himself and Roosevelt, was little short of treason.

Churchill's hostility to Bohr warped his judgement so much that his hostility crept into the agreement. "Enquiries should be made," it said, "regarding the activities of Professor Bohr and steps taken to ensure that he is responsible for no leakage of information particularly to the Russians. . . . The suggestion that the world should be informed regarding Tube Alloys, with a view to an international agreement regarding its control and use, is not accepted. The matter should continue to be regarded as of the utmost secrecy."[33] Tube alloys was the code for the atomic bomb.

This was in effect a misrepresentation of what Bohr was saying. He never at any point argued that "the world" be informed about the bomb. He focused on the Soviet Union only and on the bomb's existence only so as to avoid its possession and use being seen as coercive. In a letter Churchill dismissed the world's greatest scientist as a publicity seeker. The president was so overwhelmed by Churchill that he even denied his earlier enthusiasm for Bohr, and the follow-up meeting that Roosevelt had promised never took place.[34]

Niels Bohr ultimately failed in his attempt to convince world leaders, Roosevelt and Churchill above all, that they should open up to the Russians before the Allies had advanced so far in their preparations for a bomb that any approach would look coercive rather than co-operative. In this Bohr was proven correct.

All the post-Hiroshima and post war efforts to create an international control system crashed against the fact that the Soviets were never going to accept any arrangement where the balance of the agreement protected the Allies' advantage that then existed. Bohr had foreseen all of this. Nonetheless he acted throughout as an honest, open-minded and far-sighted member of the international community.

The third nuclear scientist who tried to prevent negative outcomes from their work in developing the atomic bomb was Eugene Paul Wigner (1902–1995). He was a humanitarian who involved himself in what he would later describe as "a horrible military weapon." He won a Nobel Prize for physics in 1963 "for his contribution theory of the atomic nucleus and the elementary particles, particularly through the discovery and application of fundamental symmetry principles." A Hungarian-American theoretical physicist, Wigner was born in Budapest to middle class non-observant Jewish parents. Wigner made a significant contribution to the project to produce atomic weapons.

After participating in the composing of the Einstein-Sziland letter to the President urging him to give the go-ahead for beginning of the Manhattan Project, Wigner led a team whose task was to design nuclear reactors that could produce weapons grade plutonium. He did this in what was then known as the Met Lab in Chicago.

Wigner was convinced that Nazi Germany was ahead of the United States in bomb development. So certain was he of this that he moved his family out of Chicago, in December 1943, when he estimated that the German head start might have given them atomic bombs by then. He thought that Germany would, logically drop the first one on the Met Lab.

Commenting on his thoughts at the time after the war was over, Wigner said:[35]

> Although none of us spoke much about it to the authorities [during this early period] . . . we did hope for another effect of the development of atomic weapons in addition to the warding off of imminent disaster. We realized that, should atomic weapons be developed, no two nations would be able to live in peace with each other unless their military forces were controlled by a common higher authority. We expected that these controls, if they were effective enough to abolish also all other forms of war. This hope was almost as strong a spur to our endeavors as was our fear of becoming victims of the enemy's atomic bombings.

Wigner hoped for world government and world peace, conditions he imagined bombs made of plutonium might enforce.

Wigner worried about proliferation of the atomic bomb to the Nazis while Szilard and Bohr worried about what would happen if the Soviets got the secret of the bomb. All three were dealing in different ways with the proliferation problem. In summer 1945 the United States became the first country to develop, test and use nuclear weapons in war. Having forced the rest of the world into the nuclear age, the politicians and policy makers had to face the unpleasant thought that their head start might not last. Even before the United States tested the weapon in the desert of New Mexico, Bohr, Szilard, Wigner, and other Manhattan Project scientists had serious misgivings. Another voice was James Franck (1882–1964), a German Jew who fled Germany in November 1933 to protest the dismissal of fellow academics by the Nazis. He first went to Nils Bohr's institute in Copenhagen and then to Johns Hopkins University in the US. Franck argued against immediately using the weapon on Japan, warning that "nuclear bombs cannot possibly remain a 'secret weapon' at the exclusive disposal of this country for more than a few years."[36] The struggle of these atomic scientists was frustrating and sometimes dangerous. "The fate of J. Robert Oppenheimer, the 'father' of the U.S. atomic bomb, is the most prominent and extreme example of the penalties a scientist could incur. . . . Oppenheimer was publicly disgraced (by being denied a security clearance) primarily because he opposed the idea of developing the hydrogen bomb."[37]

The destruction of Hiroshima and Nagasaki by single bombs disturbed and frightened the world. Americans in particular felt at risk for the first time to some annihilating attack against them, a reaction perhaps in part to having been the first to inflict such devastation on another people. A great debate ensued in the United States in the weeks and months after the war had ended and peace had been declared. There were calls to outlaw the bomb, calls to establish world government, prognostications of "one world or none."[38]

After a time of confusion about appropriate legislation, the Truman administration supported civilian control of atomic energy domestically. Internationally an Agreed Declaration with Britain and Canada concluded on November 15, 1945 that it supported action "to prevent the use of atomic energy for destructive purposes" and "to promote the . . . utilization of atomic energy for peaceful and humanitarian ends."[39]

Such action required a plan. Jimmy Byrnes as Secretary of State got the job of devising it. Byrnes appointed Dean Acheson chairman of a committee that included Groves, and the two civilian wartime leaders of the atomic bomb development, Vannever Bush and James Bryant Conant. Acheson then proceeded to create a five-person board of expert consultants. David Lilienthal, a lawyer who was head of the Tennessee Valley Authority, agreed to be its chairman and the board also included Robert Oppenheimer.

Oppenheimer came prepared having consulted with Niels Bohr at Los Alamos. They developed ideas that became the basis of the Acheson-Lilienthal Report. The work of the creation of a plan was a way to move beyond misgivings and despair about nuclear weapons. Gordon Arneson, the State Department's specialist on nuclear energy, says Oppenheimer became "the chief teacher for the Acheson-Lilienthal group."[40]

To Lilienthal, the group's work was "one of the most memorable intellectual and emotional experiences of my life."[41] Discussion intensified. Ideas came from every side-these were men of diverse backgrounds and convictions—and they debated their ideas night and day. When patience gave way to exasperation and someone proposed simply outlawing the bomb Lilienthal always waved a newspaper clipping about the Agreed Declaration to remind them that their government had already committed itself to international control of nuclear weapons.[42]

What they came to was a radical proposal. Remarkably, it managed to win their common agreement. When Bohr read it, he wrote Oppenheimer of his "deep pleasure."[43]

The Acheson-Lilienthal Report found that outlawing atomic bombs by inspection was unreliable. "Every stage in the activity, leading from raw materials to weapons, needs some sort of control." That control might work "if the element of rivalry between nations were removed." The way to do that was to assign "the intrinsically dangerous phases of the development of atomic energy to an international organization responsible to all peoples."[44]

If, for example, only this Atomic Development Authority could legally own and develop uranium ore, "then . . . not the purpose of those who mine or possess uranium ore, but the mere fact of their mining or possessing it becomes illegal, and national violation is an unambiguous danger signal of warlike purposes."

Three kinds of activity were thought to be essentially dangerous: acquiring raw materials, producing fissionable uranium and plutonium and using them to make atomic weapons. Other kinds of activity were thought to be safe: using radioactive tracer materials in medicine and science; opening small research reactors, possibly operating power reactors as well if the fuel were properly "denatured" to make converting it to an explosive more difficult. Safe activities could be licensed to states, thereby reducing the necessary scale of the Atomic Development Authority bureaucracy. This led to the crucial question of sanctions.

If a nation were to attempt to violate such an agreement as the group was proposing, how could it be punished? Would the authority maintain an army? Would it stockpile atomic bombs? No, instead the authority as it developed would spread its mines and factories around. Spreading such development around would be a way of transitioning from national to international control.[45] This remarkable idea of spreading the intrinsically dangerous mines and factories around is indistinguishable from what has come to be called nuclear proliferation, except that the agent of proliferation in the board's design would have an organ of the United Nations rather than individual states, and the technology that rated it would have been infrastructure alone rather than infrastructure and stockpiled weapons. The report notes more than once that true security is incompatible with secrecy.

In the form in which the Acheson-Lilienthal Report proposed it in March 1946, it never had a chance.

When the board was finishing its seven weeks of intense deliberation, Truman moved to appointing a conservative to head the delegation that would present the US proposal to the United Nations. Truman chose the multi-millionaire financier Bernard Baruch.

When Baruch, freshly appointed, read about the Acheson-Lilienthal Report in the newspapers he was enraged. When he learned, a day or two after it had been officially released on March28, that Acheson had said the report would be the basis for discussion at the United Nations, he was even more furious.

Baruch's first concern was that the Acheson-Lilienthal Report did not deal with the problem of enforcement. He was convinced "that world peace is impossible without the force to sanction it."[46] He would insist on "sanctions against those who violated the rules."

Baruch's view prevailed. His Baruch Plan also proposes that the United States would give up its stock of nuclear weapons only as the other nations fell into line.

Efforts to ban the use of nuclear weapons started at a time when the United States was the only nuclear power. The very first proposal emanated from the Soviet Union in 1946.[47] The Soviet Union countered the Baruch Plan with a proposal for immediate and universal disarmament without inspection. As opposed to the Baruch Plan for the creation of an International Atomic Development Authority, a draft convention prohibiting the production and use of atomic weapons was submitted by the Soviet Union to the newly established United Nations Atomic Energy Commission to which Baruch was the American delegate. They discussed the proposals until December when it voted on a plan largely modeled on the American plan, the Soviet Union and Poland abstaining. "There," writes Acheson, "the matter died." Nothing came of the effort except bad will.[48]

On September 23, 1949 Truman announced the explosion of Joe I, the first Soviet atomic bomb. This came as a great shock to the United States. It should have been obvious that the Russians would go to great lengths to break the American monopoly on nuclear weapons as soon as possible.[49]

Since the advent of the nuclear age, the security of nations has taken new dimensions unparalleled in the history of mankind. The experience of Hiroshima and Nagasaki remains as vivid examples of the destructive and indiscriminate nature of nuclear weapons. The part that nuclear weapons has also played as an instrument of foreign policy and deterrence has marked the post-World War II era. In such an atmosphere, nations have followed different routes to ensure their security. Some have found that the best way to make sure of their own security was to acquire the lethal weapons themselves. Others have joined military alliances in which they enjoyed a sort of guarantee of nuclear protection by the major nuclear ally who agreed to offer them shelter and protection. Some have even signed bilateral agreements for defense purposes with the nuclear ally. For the majority of nations, the non-aligned, security has generally been sought through efforts to achieve nuclear disarmament and arms control measures including the non-use or threat of use of nuclear weapons.[50]

The first response of the United States to the Soviet Union's demonstrated mastery of explosive fission was to approve, in October 1949, a program to increase the production of uranium and plutonium.[51] NATO's response was a massive program of rearmament, and Washington expanded its nuclear forces to preserve its "deterrent" that backed up NATO troops on the ground.[52]

Irish efforts at the United Nations to achieve a nonproliferation agreement ultimately led to the "Irish Resolution" of 1961 that laid the initial groundwork for the Non-Proliferation Treaty of 1968. The Eisenhower administration initially opposed these Irish efforts since that cut against the US policy of dispersing nuclear weapons in Europe.

In 1961 the UN General Assembly unanimously approved an Irish resolution calling on all states, particularly the nuclear powers, to conclude an international agreement to refrain from transfer or acquisition of nuclear weapons.[53]

The nuclear physicists who worked on the Manhattan Project to develop a nuclear weapon, those responsible for the Acheson-Lilienthal Report presented by Bernard Baruch at the United Nations, and President Dwight Eisenhower in his Atoms for Peace speech at the United Nations were all concerned about nuclear proliferation and an arms race between the United States and Russia. The discovery of nuclear power offered great hope and great danger for the world community if it was not dealt with wisely and in a timely fashion.

The early postwar efforts to achieve agreement on nuclear disarmament failed. After the Soviet's success in 1949, the United Kingdom in 1952, France in 1960 and the People's Republic of China in 1964 all became nuclear weapons states. The need to bring the nuclear proliferation that had begun to stop had become clear, and led to the 1968 Treaty on the Nonproliferation of Nuclear Weapons, signed in 1970, which will be the subject of the next chapter.

NOTES

1. Lifton, Robert Jay, 1967, *Death in Life Survivors of Hiroshima,* New York: Random House.
2. Lifton, *Death in Life Survivors of Hiroshima.*
3. Lifton, *Death in Life Survivors of Hiroshima,* p. 17.
4. Hachiya, M., 1955, (Warner Wells, ed. and trans.), *Hiroshima Diary*. Chapel Hill: University of North Carolina Press; In: Lifton, *Death in Life Survivors of Hiroshima,* p. 18.
5. Lifton, *Death in Life Survivors of Hiroshima,* p. 19.
6. Lifton, *Death in Life Survivors of Hiroshima,* p. 20.
7. Brodie, Bernard ed., 1946. *The Absolute Weapon: Atomic Power and World Order,* New York: Harcourt Brace, p. 21.
8. Rhodes, Richard 2013, "Why Did We Make Atomic Bombs?" James Conca Forbes https://www.forbes.com/sites/jamesconca.
9. Watson, Peter, 2018. *Fallout. Conspiracy, Cover-Up, and the Deceitful Case for the Atom Bomb A New History.* New York: Public Affairs, p. 4.
10. Watson, *Fallout. Conspiracy, Cover-Up,* p. 29.
11. Rhodes, Richard, 1988. *The Making of the Atom Bomb*. New York: Simon and Schuster, pp. 502–3.
12. Hewlett, Richard G., and Anderson, Oscar F., 1962, *The New World 1939/1946: Volume I of a History of the United States Atomic Energy Commission,* Philadelphia: Pennsylvania State University Press, p. 344, cited in Watson, *Fallout. Conspiracy, Cover-Up,* p. 321.
13. Rhodes, *The Making of the Atom Bomb,* p. 749.

14. Rhodes. Richard, 1995. *Dark Sun. The Making of the Hydrogen Bomb.* New York: Simon and Schuster; Szilard, Leo, 1961. "My Trial as a War Criminal," in: *The Voice of Dolphin and Other Stories,* New York: Simon and Schuster, pp. 75–89.
15. Reed, Thomas C. and Stillman, Danny B.2009. *The Nuclear Express. A Political History of the Bomb and its Proliferation.* Minneapolis, MN: Zenith Press, P. 10.
16. Watson, *Fallout. Conspiracy, Cover-Up,* p. 15.
17. Watson, *Fallout. Conspiracy, Cover-Up,* p.15.
18. Rhodes, Richard, 1995. *Dark Sun: The Making of the Hydrogen Bomb.* New York: Simon and Schuster, p. 206.
19. Sagdeev, Roald, "Reconstructed Account of the Conversations between Niels Bohr and Y. P. Terletsky," In Rhodes, *Dark Sun,* p. 220.
20. DeGroot, Gerard. 2006. *The Bomb. A Life.* Cambridge MA: Harvard University Press, p.69.
21. Sherwin, Martin, 1975. *A World Destroyed: The Atom Bomb and the Grand Alliance.* New York: Knopf, p. 91.
22. Watson, *Fallout. Conspiracy, Cover-Up,* p. 271.
23. Sherwin, *A World Destroyed,* p. 95.
24. Sherwin, *A World Destroyed,* p. 97.
25. Watson, *Fallout. Conspiracy, Cover-Up,* p.273.
26. Sherwin, *A World Destroyed,* p. 100.
27. Sherwin, *A World Destroyed,* p. 100.
28. Watson, *Fallout. Conspiracy, Cover-Up,* p. 284.
29. Farmelo, Graham, 2014. *Churchill's Bomb: A Hidden History of Britain's First Nuclear Weapons Program,* London: Faber & Faber, p. 262.
30. Sherwin, *A World Destroyed,* p. 109; Watson, *Fallout. Conspiracy, Cover-Up,* p. 306.
31. Ibid.
32. Ibid.
33. Watson, *Fallout. Conspiracy, Cover-Up,* p. 310.
34. Pais, Abraham. 2007. "Niels Bohr's Times," In: *Physics, Philosophy and Policy.* Oxford: Oxford University Press: 2007, p. 363.
35. Rhodes, *The Making of the Atomic Bomb,* p. 308.
36. "A Report to the Secretary of War" (The Franck Report), https://fas.org/spg/eprint/franck.html.
37. Powaski, Ronald E. 2000, *Return to Armageddon. The United States and the Nuclear Arms Race, 1981–1999.* New York: Oxford University Press, p, 7.
38. Rhodes, *Dark Sun,* p. 229.
39. Rhodes, *Dark Sun,* p. 229.
40. Rhodes, *Dark Sun,* p. 229.
41. Rhodes, *Dark Sun,* p.230.
42. Rhodes, *Dark Sun,* p. 231.
43. Rhodes, *Dark Sun,* p. 231.
44. Rhodes, *Dark Sun,* p. 231.
45. Rhodes, *Dark Sun,* p. 232.
46. Rhodes, *Dark Sun,* p 239.
47. Shaker, Mohammed I. 1980. *The Nuclear Non-Proliferation Treaty : Origin and Implementation. 1959–1979,* Vol. II, London: Oceana Publications, p. 479.
48. Shaker, *The Nuclear Non-Proliferation Treaty,* p. 240.
49. Dyer, Gwynne. 2016. *War,* New Edition, Toronto: Vintage Canada, p. 297.
50. Shaker, *The Nuclear Non-Proliferation Treaty,* p. 472.
51. Rhodes, *The Making of the Atomic Bomb,* p. 767.
52. Dyer, p. 298.
53. Miller, Nicholas I. 2018. *Stopping the Bomb. The Sources and Effectiveness of US Nonproliferation Policy.* Ithaca: Cornell University Press, p.42.

THREE

The Nuclear Non-Proliferation Treaty

On March 5, 2020, the Nuclear Nonproliferation Treaty (NPT) marked the fiftieth anniversary of its entry into force, and the twenty-fifth anniversary of its indefinite extension. It is one of the most successful security treaties in history. The NPT's norms and monitoring procedures are an important reason why only thirteen states have ever held nuclear weapons and four of these—South Africa, Belarus, Kazakhstan, and Ukraine—have given them up. There are at present only nine nuclear states in the world today.

The NPT is often discussed along with Resolution 255 of the United Nations Security Council. Resolution 255 was adopted on June 19, 1968, and remains the only global solution reached in connection with the NPT.[1] Resolution 255 appeared after a large number of states began to subscribe to the Treaty on the Non-Proliferation of Nuclear Weapons. The Security Council recognized that aggression with nuclear weapons or the threat of it against a non-nuclear weapon state would create a situation in which the Security Council and its nuclear-weapon state members would have to act immediately in accordance with their obligations under the UN Charter. The Council also welcomed the intention expressed by certain states that they would provide or support immediate assistance to a non-nuclear weapon state who had signed the NPT if and when it was a victim of an act or the object of a threat to use nuclear weapons. In this way it reaffirmed the inherent right to individual and collective self-defense.

Even with the almost universal coverage of the NPT and the global scope of Resolution 255 the world continues to live with the serious threat of global nuclear annihilation. This was succinctly expressed by the axiom of the Canberra Commission in 1996, which still holds true: "As long as any state has nuclear weapons, others will seek to acquire them." The opening statement of the Commission consists of plain truths which we all need to keep in mind:[2]

> The destructiveness of nuclear weapons is immense. Any use would be catastrophic.
>
> Nuclear weapons pose an intolerable threat to all humanity and its habitat, yet tens of thousands remain in arsenals built up at an extraordinary time of deep antagonism. That time has passed, yet assertions of their utility continue.
>
> Nuclear weapons are held by a handful of states which insist that these weapons provide unique security benefits, and yet reserve uniquely to themselves the right to own them. This situation is highly discriminatory and thus unstable; it cannot be sustained. The possession of nuclear weapons by any state is a constant stimulus to other states to acquire them.

Early in his presidency in the spring of 2009, Barack Obama made a significant commitment on behalf of the United States to seek the peace and security of a world without nuclear weapons in his speech in Prague in the Czech Republic.[3]

> The existence of thousands of nuclear weapons is the most dangerous legacy of the Cold War. No nuclear war was fought between the United States and the Soviet Union, but generations lived with the knowledge that their world could be erased in a single flash of light. Cities like Prague that existed for centuries, that embodied the beauty and the talent of so much of humanity, would have ceased to exist.
>
> Today, the Cold War has disappeared but thousands of those weapons have not. In a strange turn of history, the threat of global nuclear war has gone down, but the risk of a nuclear attack has gone up. More nations have acquired these weapons. Testing has continued. Black market trade in nuclear materials abounds. The technology to build a bomb has spread. Terrorists are determined to buy, build or steal one. Our efforts to contain these dangers are centered on a global non-proliferation regime, but as more people and nations break the rules, we could reach the point where the center cannot hold.

Obama then availed himself of the axiom regarding proliferation in the Canberra Declaration to transform it into an assertion of personal and American responsibility.

> Some argue that the spread of these weapons cannot be stopped, cannot be checked—that we are destined to live in a world where more nations and more people possess the ultimate tools of destruction. Such fatalism is a deadly adversary, for if we believe that the spread of nuclear weapons is inevitable, then in some way we are admitting to ourselves that the use of nuclear weapons is inevitable.
>
> Just as we stood for freedom in the 20th century, we must stand together for the right of people everywhere to live free from fear in the 21st century. And as the only nuclear power to have used a nuclear

> weapon, the United States has a moral responsibility to act. We cannot succeed in this endeavor alone, but we can lead it, we can start it.
>
> So today, I state clearly and with conviction America's commitment to seek the peace and security of a world without nuclear weapons.

To appreciate the significance of President Obama's pledge to ensure that the NPT will continue and that the United States will honor its commitment to seek a world of peace and security without nuclear weapons it is necessary to revisit the history of the creation of the NPT.

In the immediate post-war period both the executive and the legislative branches of the United States government rapidly moved toward the view that America's monopoly on atomic weapons should be preserved as long as possible. In order to achieve this, the United States would retain national control of all nuclear weapons and deny information on nuclear technology to all other nations. Attainment of these goals was complicated by the fact that Britain had played a significant role in developing the atom bomb and Churchill and Roosevelt had agreed that the use should be a matter of joint Anglo-American decision. Truman was working toward an accommodation with Britain when an unexpected turn of events took place.

The actions of the executive branch on sharing nuclear information was preempted by Congress who, in the summer of 1946, passed a bill sponsored by Senator Brian McMahon forbidding the United States government from sharing information about nuclear technology with any other power. Until its repeal in 1958 Anglo-American nuclear relations developed within the framework of the McMahon Act.[4]

With the Soviets' rejection of the Baruch UN Plan in 1946, and their explosion of a nuclear device in 1949, Washington began to fear that Moscow might use its nuclear force to destroy the United States.[5] Baruch had noted in his UN address that "no nation would think of starting a war with only one bomb."[6]

A different early estimate of the number of warheads needed by the US to win a nuclear war can be found in the Franck Report. This was prepared under Arthur Compton, director of the Manhattan Project's plutonium production efforts and was sent to the US Secretary of War Henry Stimson. Its purpose was to explore how the development of atomic energy might be controlled after the war.[7] An estimate made in the Franck Report was that between 50 and 170 Hiroshima sized bombs would be needed.

Beyond possibly developing a pre-emptive strike option that might be triggered by the first signs of a Soviet strike President Dwight Eisenhower began "to search around for the kind of idea that could bring the world to look at the atomic problem in a broad and intelligent way and still escape the impasse to action created by Russia's intransigence in the matter of mutual or neutral inspection of resources."[8]

In his "Atoms for Peace" speech in December 1953, Eisenhower offered the suggestion that governments involved in developing atomic weapons should start to make contributions from their stockpiles of normal uranium and fissionable materials to an international atomic energy agency. The atomic energy agency could be made responsible for the impounding, storage and protection of the contributed fissionable and other materials. The important responsibility of this atomic energy agency would be to devise methods whereby this fissionable material would be transformed to serve the peaceful ambitions of humanity. Experts would be mobilized to make use of atomic energy for the needs of agriculture, medicine and other peaceful undertakings. A special purpose would be to provide abundant electrical energy in the power-starved areas of the world.[9] Unfortunately, in Eisenhower's Atoms for Peace speech there was no explanation as to how the IAEA might prevent the nuclear reactors and relocated technology it was planning to share with the world's non-weapons states from being used to produce nuclear weapons-usable materials. Later in the book, we will show how India has proven how difficult it is to prevent the military application of nuclear technology and materials designed for purely civilian purposes.

Alongside the "Atoms for Peace" ideas of Eisenhower, another important concept developing in this period was "atom-free zones." As early as 1952 and especially from 1958, the Soviets began to show a decided interest in the establishment of various atom-free zones.

In 1961, the UN General Assembly first adopted resolution 1653 (XVI) originally proposed by Ethiopia, which called upon member states to consider and respect the continent of Africa as a de-nuclearized zone. In 1964, the Organization of African Unity (OAU) issued the Declaration on the De-Nuclearization of Africa, which was subsequently endorsed by the United Nations General Assembly. The Africa Nuclear Weapons Free Zone (ANWFZ), the Treaty of Pelindaba, was signed in Cairo on April 11, 1996, by forty-seven of the continent's states.

In addition to supporting atom free zones such as the ANWFZ, the Soviets also pushed hard for a nuclear test ban. But at the same time there was also a mounting concern on the part of the Soviet Union with respect to the whole question of proliferation. This concern about proliferation was primarily in terms of the Sino-Soviet relationship.[10]

The USSR had signed an agreement with the Chinese on October 15, 1957, concerning the transfer of nuclear technology and sample materials from the Soviet Union to China. This agreement was unilaterally abrogated by the Soviets on June 20, 1959. It was abrogated in the wake of the Taiwan Straits crisis, when the People's Republic of China attempted to wrest control of the offshore island of Quemoy from Taiwan in the summer of 1958. The Soviets, alerted by the enormous seriousness of the situation, were made aware of the dangerous prospect that in some future crisis, in which China possessed its own atomic bombs, Moscow might be dragged into a catalytic nuclear confrontation with the United States because of some irresponsibility of its erstwhile ally[11] of China. The experience of this crisis marked a significant watershed not only in the deterioration of the Sino-Soviet relationship but also in the evolution of Moscow's opposition to nuclear proliferation.[12]

By the summer of 1968, the five victors of World War II, now permanent members of the UN Security Council, had not only joined the nuclear club, but they had tested (or were about to test) two-stage thermonuclear weapons. They well understood that Israel constituted a de-facto sixth nuclear weapon state, since many had aided and abetted that situation, and they all recognized that the losers and neutral bystanders of World War II such as West Germany were perfectly capable of going nuclear on short notice if they decided to do so. In addition, regional industrial states with grudges and fears were considering the nuclear option, and two principal nations of south Asia were enjoying covert benefits from one or more of the Big Five.[13] It became the collective wisdom of American, British and Soviet academics that something had to be done to cap the flow of nuclear weapon technology or the actual transfer of such weapons before matters ran out of control. On a parallel track, the nonaligned movement had long advocated the total abolition of nuclear weapons.[14]

This idea was echoed by France which eventually became a member of the "Big Five" nuclear weapons possessing countries.[15] In a November 1964 speech before the French National Assembly, the French Minister of Europe and Foreign Affairs, Maurice Couve de Murville, asked the following question: "How can one claim to deny to others, unless they voluntarily renounce it for themselves, what one permits oneself? Experience proves, moreover, that this is never accomplished, except through force, which has never to my knowledge been considered. The solution, the only solution, is the total destruction of the existing arsenals and the banning of all new manufacture, both under effective control."[16] But in the event that the abolition of nuclear weapons was not immediately feasible, serious efforts to prevent their proliferation could and should be undertaken.

In the end, France successively tested fusion weapons of mass destruction in the desert of Algeria on August 23, 1968. It has conducted 210 above ground tests and is thought to have three hundred warheads at the present time.

It was not, however the crisis between the Soviet Union and the People's Republic of China that was the true beginning of the process leading to the Nuclear Nonproliferation Treaty. Fundamental to the Nuclear Non-Proliferation agreement in the United Nations was the initiative of one of its member states, namely Ireland. In 1959, upon Ireland's request the question of the "prevention of the wider dissemination of nuclear weapons" was included in the agenda of the General Assembly.[17]

The first proposals by the Irish delegation in 1958 were not well received by the United States as well as other NATO members. In part this can be explained by the situation that resulted after the successful launching by the USSR of the "Sputnik" space craft on October 4, 1957, and the subsequent emplacement of Soviet missiles aimed at Europe, all of which had shaken the confidence of both the United States and Western Europe in the effectiveness of the American deterrent system. As a result, the United States decided to deploy nuclear delivery systems including intermediate range ballistic missiles in Europe, and then entered into a series of bilateral nuclear stockpile arrangements with over half of its NATO allies. Basically, these arrangements provided the authority for the release of American nuclear weapons stored in Europe to the appropriate NATO commanders in the event of hostilities. Therefore, a non-proliferation resolution would have been a psychological barrier to the type of nuclear arrangements the United States was negotiating. In other words, the United States was putting flexibility in the use of nuclear weapons above the international efforts to erect barriers against the acquisition of nuclear weapons.[18]

Irish Foreign Minister Frank Aiken offered a draft resolution concerning the "Further Dissemination of Nuclear Weapons" before the First Committee of the UN General Assembly on October 17, 1958. The resolution was quite modest, recognizing as it did that "an increase in the number of states possessing nuclear weapons may occur aggravating international tensions" and making disarmament "more difficult." It went on to recommend that the General Assembly establish an ad hoc committee to study the dangers inherent in the further dissemination of nuclear weapons. On October 31, 1958, Aiken withdrew the resolution when it became clear that no NATO nation was yet ready to endorse the initiative.[19]

The effort of 1958 was renewed in 1959. It was "hardly realistic," Aiken argued, to expect any "early agreement on the abolition of nuclear weapons. But what we can do," Aiken argued, "is to reduce the risks which the spread of these weapons involves for this generation, and not to hand on to our children a problem even more difficult to solve than that with which dissemination of nuclear weapons" was included as a separate item in the agenda of the fourteenth session of the General Assembly.

There were twin structures in Aiken's 1959 proposal. Weapons states should not furnish nuclear weapons to non-weapons states, and non-weapons states should refrain from trying to acquire them.[20] The key reason for demanding a nonproliferation treaty, after all, was to address the proliferation threat without having to wait for the advent of complete nuclear disarmament.[21]

As a result of this initiative and the importance of the topic, a Ten-Nation Disarmament Committee was set up in 1959 and by agreement reached among the governments of France, the USSR, the United Kingdom and the United States to consider disarmament matters. The Irish delegation submitted to the First Committee of the General Assembly a draft resolution suggesting that the Ten Nation Committee should, in the course of its deliberations on disarmament consider also an appropriate means by which the danger of an increase in the number of states possessing nuclear weapons be averted "including the feasibility of an international agreement, subject to inspection and control, whereby the Powers producing nuclear weapons would refrain from handing over the control of such weapons to any nation not possessing them and whereby the Powers not possessing such weapons would refrain from manufacturing them."[22]

The draft resolution of 1959 incorporated the second paragraph of the withdrawn Irish resolution of 1958 and went further than the previous Irish proposals in suggesting and singling out an international agreement as one means to check the proliferation problem.[23]

The new Irish draft was welcomed by most delegations. The United States, which was displeased with the 1958 Irish proposals, supported the 1959 draft resolution. By that time certain developments had moved the United States in the direction of an active non-proliferation policy. The White House realized that Congress would not consent to a major extension of nuclear weapons cooperation to any country but Great Britain. The United States was also beginning to suspect that it had overreacted to "Sputnik" and that it was capable of handling the Soviet missiles without the aid of allies. It also came to the conclusion that its "atoms of peace" approach was actually helping countries to develop at least a threshold nuclear capability.[24]

In the end, the Soviet Union did not support the Irish draft resolution of 1959 because the draft did not prohibit states from having nuclear weapons outside their own territory, outlawing such weapons and destroying stockpiles on foreign bases. For the Soviet Union those problems had to be resolved if the wider dissemination of nuclear weapons was to be prevented at all.[25]

The Irish draft was adopted by the First Committee on November 16, 1959, and a few days later by the General Assembly on November 20 by sixty-eight votes in favor, none against, and twelve abstentions (including the USSR).

The importance of that first Irish resolution lies primarily in the recognition by the General Assembly of an existing danger in the proliferation of nuclear weapons. That concept of danger was maintained during the different phases of the NPT negotiations.[26]

It was only on December 4, 1961, that the General Assembly upon yet another Irish initiative[27] was able to adopt unanimously a resolution on the "prevention of wider dissemination of nuclear weapons,"[28] a resolution which has come to be widely known and referred to in disarmament and arms control negotiation jargon as the "Irish Resolution."

The resolution calls upon all States, and in particular upon the States at present possessing nuclear weapons, to use their best efforts to secure the conclusion of an international agreement containing provisions under which the nuclear states would undertake to refrain from relinquishing control of nuclear weapons and from transmitting the information necessary for their manufacture to states not possessing such weapons and provisions under which States not possessing nuclear weapons would undertake not to manufacture or otherwise acquire control of such weapons. The resolution also urges all states to co-operate to achieve such ends, Under the NPT, nuclear countries have the obligation to stem vertical proliferation (increasing stockpiles within the country) and horizontal proliferation (sharing the technology with others).

One should not confuse America's efforts to promote negotiation of the Nuclear Nonproliferation Treaty (NPT) with the Atoms for Peace Program. Both efforts encouraged nuclear technology transfers and international atomic energy safeguards. Both called on nations to abandon efforts to acquire nuclear weapons.

Yet what distinguishes these non-proliferation initiatives from one another—including their differing views of what strategic developments were most worrisome—is far more important than any similarity between the two. Where the Atoms for Peace was geared to reduce the threat supposedly posed by arsenals large enough to destroy most of America's major cities, the NPT was directed against catalytic[29] and accidental nuclear wars that could arise either from the further spread of nuclear weapons or from the build-up of the superpowers arsenals.[30]

Those who first backed the treaty's negotiation were primarily concerned with how horizontal proliferation (i.e., the spread of nuclear weapons to other nations) might intentionally or accidentally spark (or catalyze) nuclear wars between the superpowers or smaller states. This gives rise to the treaty's prohibitions and safeguards against sharing or accepting nuclear weapons technology (Articles I, II, and III of the NPT).

A decade later, however, when the treaty's negotiation was nearly complete, this was not the key concern. Instead, most of the treaty's negotiators were preoccupied with how growing vertical proliferation, i.e., the quantitative and qualitative improvement of the superpowers' strategic arsenals) might lead to unauthorized or accidental nuclear wars and prompt other nations to go nuclear. This gave rise to the treaty's provisions that the nuclear weapons powers shared their nuclear technology with the world's non-weapons states without reservation (Articles IV and V) and that they work toward total disarmament (Article VI).[31] If the non-nuclear weapons states were to feel threatened, they should be free to acquire nuclear weapons of their own (Article X). In the late 1950s, experts who worried about nuclear proliferation were persuaded that US authority over these weapons was less than complete.[32]

William Foster, director of the US Arms Control and Disarmament Agency (ACDA) reported back to Washington after the 1965 session of the UN Disarmament Commission on the "growing resistance" of non-aligned states to non-proliferation arrangements "that do not impose any obligations on the nuclear powers to limit their arms."[33]

On July 1, 1968, the resulting treaty on the Nonproliferation of Nuclear Weapons (NPT) was enacted by the United States, the United Kingdom and the Soviet Union along with fifty-nine other countries. The People's Republic of China and France did not sign, but they implied that they would do so when their nuclear weapons programs had matured and when they had therefore caught up with the Big Three as well as satisfied their nuclear technology transfer obligations to allies and customers.[34]

The NPT entered into force on March 5, 1970, with the deposit of ratification documents at the UN. There was, after the end of the Cold War, a small stampede to join the Nuclear Non-Proliferation Treaty. In the following years, virtually every other nation has joined the NPT regime.[35] Most importantly, The People's Republic of China and France entered the treaty on March 9, 1992, and on August 3 of that year, respectively. South Africa, in a remarkable rollback move, dismantled its secret nuclear arsenal, and acceded to the NPT in 1995 and 1998 respectively.[36] The disassembly of the Soviet Union in 1991 created four nuclear weapons states where there had been only one previously. But soon thereafter Belarus, Kazakhstan, and the Ukraine transferred all the Soviet nukes in their territories back to the Russian Federation and then acceded to the treaty. More about this significant development in nuclear non-proliferation will be provided in a following chapter of the book.

At present, 187 states are parties to the NPT. Cuba was the last to sign, in 2002, while only Israel, India, Pakistan, and Southern Sudan have failed to join. North Korea withdrew on January 10, 2003. Taiwan, although no longer a UN member, has pledged its continuing observance of the treaty provisions.

The NPT obligates the five acknowledged nuclear weapons states (China, France, Russia, the United Kingdom, and the United States) not to transfer nuclear weapons, other nuclear explosive devices, or the associated technology to any non-nuclear state. For their part, the non-nuclear weapons states agree to foreswear the acquisition or production of nuclear weapons or nuclear explosive devices and to accept safeguards to detect the diversion of nuclear materials from peaceful activities, such as power generation. The International Atomic Energy Agency (IAEA) with its headquarters in Vienna, was established to supervise these safeguards and to conduct inspections, but only with the concurrence of the host country. Implicit in this agreement was the intent of the major powers to provide a nuclear umbrella over their allies, removing the incentive for them to go nuclear and for those Big Five to refrain from the use, or the threat of use, of nuclear weapons on non-nuclear states unless such states were acting in alliance with a nuclear weapon state. In recent years, however, the United States, the United Kingdom, and France have all noted the possibility of using nuclear weapons in response to "nonconventional attack" by "rogue states" or their terrorist agents.[37]

The first indications of the NPT's limitations came in May 1974 when India exploded a "peaceful" nuclear device that employed "civilian" US, Canadian, and Western European reprocessing and heavy water technology and hardware.[38] Although India had not signed the NPT, this event, perhaps more than any other, raised doubts about the adequacy of merely securing peaceful end-use pledges in exchange for supplying sensitive nuclear technology—an approach many NPT proponents had hoped would be sufficient.[39]

Over the years, the NPT has come to be seen by the Third World States as a conspiracy of the nuclear "haves" to keep the nuclear "have nots" in their place. Abdul Qadeer Khan, mastermind of the Pakistani nuclear franchising industry, referred to the NPT in 1990 (before his proliferation efforts had been exposed) as "highly discriminatory in nature; it allows the nuclear weapons powers to accumulate and enhance their stockpiles of nuclear weapons while putting severe restrictions on even the peaceful utilization of nuclear power by the have-nots."[40]

Once negotiations on a nuclear non-proliferation treaty got under way in the mid-1960s, discussions of possible non-use commitments by the nuclear powers shifted. During the negotiations the non-aligned states put pressure on the nuclear powers for non-use assurances. If the non-nuclear weapon states were being asked to foreswear nuclear weapons, the non-aligned argued, the least the nuclear weapon states could do was to agree not to use or threaten to use nuclear arms against non-nuclear weapon states.[41] In late October 1966, the non-aligned states sought such assurances in a draft resolution to the General Assembly's First Committee, sponsored by thirty-two Latin American, Asian, and African states.[42] The United States and the Soviet Union seriously, but unsuccessfully, sought a formula on what came to be known as "negative security assurances" or non-use pledges to non-nuclear states.[43] US leaders were willing to accept some limitations on their use of nuclear weapons. Because of unease among their allies, the United States soon ended further discussions on "positive security assurances," that is, promises to come to the aid of a non-nuclear weapon state threatened by a nuclear attack.

This argument has its root in Article VI of the treaty, which obligates the nuclear weapons states to liquidate their nuclear stockpiles and pursue complete disarmament. The non-nuclear weapon states see no sign that this is happening. They also take offence at the nuclear arming of NATO, which they consider a violation of the treaty. Nominal US "custody," as described above, is a political fig-leaf used to fend off this argument, and it is one reason why the installation of Permissive Action Links (PALs) within the NATO-based weapons was so important.

The definition of a PAL is a device included in or attached to a nuclear weapons system to preclude its arming and/or launching until the insertion of a prescribed discrete code or combination.[44]

Unfortunately, the NPT has one giant loophole. Article IV gives each non-nuclear weapon state the "inalienable right" to pursue nuclear energy for the generation of power. This entitles such states to enrich natural uranium (0.7 percent U-235) to fuel grade (3 percent U235). From there it is a short step to weapons grade highly enriched uranium (HEU: over 90 percent U-235). A power reactor produces neutrons that are indifferent to their ultimate use. They can breed plutonium, generate power, or both.

The NPT was set to expire after twenty-five years (in 1995), but in that year the parties to the NPT met in New York and agreed to extend the treaty indefinitely with review conferences every five years. By then, however, the challenge of nuclear proliferation had taken on a new dimension: the recently discovered covert programs in North Korea and Iraq. The latter country had successfully circumvented IAEA safeguards because the IAEA could only examine "declared" nuclear facilities.

Despite these serious challenges the 1995 NPT review conference represented the high point of the decade in terms of the efforts to reduce nuclear dangers. Indications of this progress include the steady expansion of nuclear weapons free zones, a renewed global movement for the abolition of nuclear weapons and the 1996 World Court advisory opinion on the legality of the use of nuclear weapons. These developments were driven primarily by non-governmental organizations (NGOs) in collaboration with non-nuclear weapons states, often over the objections of the nuclear powers. The efforts managed to put the nuclear weapons powers on the defensive, while creating the possibility for the non-nuclear weapons states to extract more explicit commitments to disarmament at the 2000 NPT review conference.[45]

Since 1993, UN personnel have been working on amendments to the NPT intended to constrain illicit nuclear activities, broad environmental sampling systems, no-notice inspections, remote monitoring equipment, and streamlined administrative procedures.

Given the difficulty of amending an existing treaty, the UN member states, in May 1997, adopted a model Additional Protocol to the NPT in the hope that the treaty signatories would voluntarily subscribe to this agreement. The Big Five signed at once. The United States led the way, in June 1998 and even France and the United Kingdom following in September of that year. China signed in December 1998. Russia completed the list in May 2000, just as the next NPT review conference was getting underway.

That conference, again held in New York, produced an "unequivocal commitment by the nuclear weapons states to the ultimate goal of complete disarmament under strict and effective international controls."[46] They also endorsed the Comprehensive Test Ban treaty, preservation of the ABM treaty, and the timely conclusion of safeguards agreements with the IAEA: that is, the Additional Protocol.

The review conference of 2005, again held at the UN, produced stark disagreement between the nuclear and non-nuclear weapons states. The US Secretary of State did not attend, it took three weeks to agree an agenda, and there was no final statement. The conference chairman described the event as a "failure."[47] Most of the four-week session was spent in confrontation. The non-nuclear weapon states demanded action by the nuclear weapon states on disarmament and ratification of the Comprehensive Test ban Treaty. Across the table, the United States and the European powers focused on the proliferation activities of North Korea and Iran, seeking sanctions and penalties for their destabilizing behavior. The treaty loopholes went unaddressed, the IAEA was left to do the best it could, constrained by its limited inspection and negligible enforcement powers. By then, perhaps half the member states had signed and put into effect the Additional Protocol, but that list did not, and does not now, include the most worrisome Middle Eastern and South Asian states. India, Israel, Pakistan and Southern Sudan are the only states that have failed to sign the Nonproliferation Treaty. Israel's refusal to do so has become the touchstone of Arab recalcitrance.[48] Egypt, Algeria and Saudi Arabia have rejected the Additional Protocol. Iran signed in December 2003 but has since withdrawn. Pakistan and India remain outside the entire nuclear-control regime. More will be said about this later in the book. A reformed Libya is the only major Arab petro-state to sign both the NPT and the Additional Protocol.

The Non-Proliferation Treaty was signed and put into effect at the height of the Cold War. However, the post-Cold War era might be even more hazardous than the hopes engendered by the release from Communist authoritarianism at the end of the millennium. While the risk of global nuclear war has gone down, as President Obama argued, the risk of a nuclear attack has gone up.

A serious problem is terrorism. The same dilemma exists with terrorism as with nuclear weapons: the impossibility of defense. "The detonation of a terrorist device in an American city is inevitable if the US continues on its present course." These are the words of Graham Allison, founding dean of the Kennedy School at Harvard and a former assistant secretary of defense.[49] Such a result could be achieved using an amateur device made from stolen Soviet-era fissionable material, by a weapon purchased from North Korea, or by a gadget assembled from Pakistani components. Nuclear Non-proliferation must take account of the risks involved with nuclear devices that could fall into the wrong hands. The collapse of the former Soviet Union in 1991 was a moment of serious global risk when the nuclear weapons in the Ukraine, Belarus, and Kazakhstan were no longer under the control of Russia. How this risk was averted is an important moment in the non-proliferation responsibilities of the NPT signatories, and will be the subject of the eighth chapter of this book.

NOTES

1. Shaker, Mohamed, 1980. *The Nuclear Non-Proliferation Treaty: Origins and Implementation 1959–1979,* London; New York: Oceana, 3 Volumes, Vol. II, p. 479.
2. Canberra Commission on the Elimination of Nuclear Weapons (1996). Report of the Canberra Commission on the Elimination of Nuclear Weapons: Canberra, Australia, Commonwealth of Australia.
3. Rhodes, Richard, 2010. *The Twilight of the Bombs Recent Challenges, New Dangers, and the Prospects for a World without Nuclear Weapons.* New York: Alfred A. Knopf, pp. 297-298.
4. Ball, S. J. 1995. "Military Nuclear Relations between the United States and Great Britain under the terms of the McMahon Act 1946–1958," *The Historical Journal,* Vol. 38, Issue 2, June 1995, p. 440.
5. Sokoloski, Henry D. 2001. *Best of Intentions: America's Campaign against Strategic Weapons.* Westport, Conneticut Praeger, p. 25.
6. The Baruch Plan, 1960. "Statement by the United States Representative (Baruch) to the United Nations Atomic Energy Commission, June 14, 1946" Reprinted in US Department of State, Documents on Disarmament, 1945–1959, Vol. 1, Washington, DC: USGPO, p. 12.
7. Sokoloski, *Best of Intentions,* p. 13.
8. Eisenhower, Dwight D. 1963. *Mandate for Change.* Garden City NY: Doubleday, p.252.
9. Address of Mr. Dwight D. Eisenhower, President of the United States to the 470 meeting of the Plenary Session of the United Nations General Assembly, Tues. 8 December 1953 https://www.iaea.org/about/history/atoms-for-peace-speech.
10. Sokoloski, *Best of Intentions,* p. 11.
11. Halperin, Morton. 1965. *China and the Bomb,* London: Pall Mall Press, pp. 78–82.
12. Lambeth, Benjamin St., 1970. "Nuclear Proliferation and Soviet Arms Control Policy," *Orbis* Vol. XIV, No. 2, Summer 1970, pp. 300, 309–11.
13. National Planning Association, 1960. *The Nth Country Problem and Arms Control,* Washington, D.C.: National Planning Association, Special Project Committee on Security through Arms Control, Pamphlet No. 108, Jan. 1960.
14. Reed, Thomas C., and Stillman, Danny B. 2009. *The Nuclear Express: A Political History of the Bomb and its Proliferation.* Minneapolis, MN, p. 143.
15. Bader, William B. 1968. *The United States and the Spread of Nuclear Weapons* New York: Pegasus, p. 16.
16. Ambassade de France, Servicede Presse et d'Information. 1964. Speeches and Press Conferences, No. 210, November 3, 1964, p. 3.
17. GAOR, 1959. 14th Session, Anns. a.i. 67, Docs. A/4125, 18 June 1959 and A/4286, 18th Nov. 1959, para 5.
18. Bader, *The United States,* pp. 38–40.
19. GAOR, 1958. 13th Sess., Anns., a.i. 64, 70 and 72, Doc. A/C.1/L.206, 17 Oct. 1958.
20. GAOR, 13th Sess., p. 43.
21. Ibid., p. 44.
22. GAOR, 14th Sess. Op.cit.
23. Sokoloski, *Best of Intentions,* p. 13.

24. Ibid., p. 14.
25. Ibid., p. 16.
26. Ibid., p. 17.
27. GAOR, 1961. 16th Sess., Anns (Vol. iii) a.i. 81, Doc. A/4845, 16 Aug. 1961.
28. GAO Res. 1961. 1665 (XVI) 4 Dec 1961.
29. Kobe, Donald H. 1962. "A Theory of Catalytic War," *The Journal of Conflict Resolution*, Vol. VI, No. 2, June 1962, pp. 125–42.
30. Sokoloski, p. 39.
31. Sokoloski, *Best of Intentions*, p. 40.
32. Ibid., p. 41.
33. Report by the Director of ADCA (Foster), 1965. Summary of Session of UN Disarmament Commission, June 18, 1965, FRUS 1964–68, 11, p. 215.
34. Reed and Stillman, *The Nuclear Express*, p. 143.
35. Tannenwald, Nina, 2007. *The Nuclear Taboo: The United States and the Non-use of Nuclear Weapons since 1945*. New York: Cambridge University Press, p. 330.
36. Albright, David. 1994. "South Africa and the Affordable Bomb," *Bulletin of the Atomic Scientists*, Vol. 50, No. 4, pp. 37-47.
37. Albright, "South Africa and the Affordable Bomb," p. 144.
38. Sokoloski, *Best of Intentions*, p.56.
39. Wohlstetter, Albert et al., 1977. "The Military Potential of Civilian Nuclear Energy," *Minerva* Vol. 15, Issue 3, (Autumn-Winter 1977), pp. 387–38, at p. 387.
40. Wohlstetter, "The Military Potential of Civilian Nuclear Energy."
41. Willrich, Mason. 1965. "No First Use of Nuclear Weapons: An Assessment," *Orbis*, Vol. 9, No. 2 (Summer 1965), pp 299-315, at 314.
42. UN Doc. A/C.1/L.371, Oct. 27, 1966.
43. Tannenwald, *The Nuclear Taboo*, p. 248.
44. US Department of Energy, Washington, DC, https://www.directives.doe.gov.
45. Tannenwald, *The Nuclear Taboo*, p. 347.
46. Arms Control Association, 2000, NPT Conference Final Document, https://armscontrol.org/act/2000_06/docjun.
47. Arms Control Association. The 2005 NPT Review Conference: Can it Meet the Nuclear Challenge? https://www.armscontrol.org/act/2005_04/duPreez.
48. Reed and Stillman, *The Nuclear Express*, p. 146.
49. Ibid., p. 4.

FOUR

The Risk of a Regional Nuclear War between India and Pakistan

Catastrophes can change the course of history. The list of human-caused catastrophic risks is familiar: nuclear weapons, global warming and other ecological disruptions, pandemics from pathogens engineered by biotechnology, runaway artificial technology, to mention a few.

For most catastrophe scenarios, the ultimate effect on the human species is hard to predict. Even if some humans survive, humanity as a whole may never again regain the advanced civilization we now enjoy.[1]

An all-out nuclear war that involved all of the world's nuclear armed countries: China, France, India, Israel, North Korea, Pakistan, Russia, the United Kingdom and the USA would constitute a global climatic and environmental catastrophe.[2] This is not a probable scenario at the present time. A more likely eventuality would only involve Russia and the US, which together hold over 90 percent of the global nuclear arsenal.

Even in the worst case, much of the world may be spared from immediate destruction. Africa and Latin America in particular have many countries that are neither close allies nor adversaries of any of the nuclear armed countries. Residents of these countries would presumably survive the initial nuclear explosions, as would people who live in the targeted countries but away from the cities and military sites that get bombed.

The survivors' world would, however, be very different. In addition to the social and political turmoil they would also lose many important entry points into the global economy. Many global supply chains are designed to be highly efficient under normal conditions but are fragile to even small disruptions—and this disruption would not be small. Within weeks or even days, communities all over the world could face shortages of consumer goods, replacement parts for critical industrial infrastructure, and other basic economic needs.

Soon after, the global environmental effects would begin. Nuclear explosions are so powerful that they can send the dust and ash from burning cities all the way into the stratosphere. The primary destructive effect of nuclear weapons is mass fire. The highly combustible South Asian megacities, which are likely targets, would send smoke from mass fires into the stratosphere, which would drift around the world, and darken and chill the earth for a decade, reducing the global average temperature below those of the European Little Ice Age of the sixteenth to mid-nineteenth centuries, when crops failed and millions starved.

The stratosphere is the second layer of the atmosphere located 7 km (5 miles) above the surface of the poles and 20 km (12 miles) at the equator. Smoke and dust there would spread around the world within a few months and stay aloft for a few years. While aloft, they would block the incoming sunlight, cooling the surface and reducing precipitation, all of which is bad for agriculture.

Recent research finds significant potential nuclear winter effects from an India-Pakistan nuclear war involving one hundred weapons of 20 KT yield dropped on major cities.[3] Even a "small" regional war, such as one between India and Pakistan, in other words, leave aside a global international nuclear war, would be an unmitigated disaster for the entire world. A nuclear war is not just a nuclear war: it is also an economic recession and an agricultural failure.[4] A regional nuclear conflict using one hundred Hiroshima-size (15 KT) nuclear weapons, only 0.03 percent of the explosive power of the current global arsenal, would produce climate change unprecedented in human history.[5]

Let us now turn our attention to the danger of a regional nuclear war between India and Pakistan. The root cause is the post-independence fate of the state of Kashmir. Indian and Pakistani leaders have for decades tied their respective nation's identities to not relinquishing Kashmir to the other.

When India and Pakistan became independent in 1947, the princely state of Jammu and Kashmir consisted of a majority Muslim population but was ruled by a Hindu maharaja. The maharaja chose to accede neither to India nor to Pakistan at the time of partition, and the English governor-general, Lord Mountbatten, could not resolve the matter. In the fall of 1947, five thousand tribesmen from Pakistan's northwest Frontier Province captured several towns in Jammu and Kashmir. To obtain military assistance from India to repel the invasion, the maharajah had to sign an Instrument of Accession to India, which he did on October 26, 1947. While India argued with some cause that the accession was legal, Pakistan disputed it. Pakistan insisted that India provide an assurance that "the question of accession should be decided in accordance with the wishes of the people of the state [and this could be] settled by a reference to the people." This was taken to mean a plebiscite, despite later Indian protestations that no such specific mechanism was agreed on to determine the wishes of the people of Jammu and Kashmir.[6]

Events continued with Pakistan sending regular military into Jammu and Kashmir to solidify its hold on the western part of the state. India answered this with its own troops, and war ensued. In the end, the United Nations brokered a truce to put a stop to the fighting. Two resolutions, one of August 13, 1948, the second of January 6, 1949, called for a cease-fire, a truce and a plebiscite. The cease-fire went into effect as of January 1, 1949. The truce has not been formally implemented up to the present time. Nor has a plebiscite been held.[7]

India's essence as a secular state has "required" the integration of Muslim-majority Jammu and Kashmir to prove that Muslims could fare well in India. Pakistan's identity as an Islamic state has "required" that all major conglomerations of Muslims in South Asia live within it, a contention India has rejected from the beginning.

In 1960, the United States Atomic Energy Commission was encouraging international enthusiasm for peaceful nuclear explosions (PNEs) for use on massive engineering projects such as canal excavations. The US nuclear establishment saw PNEs as a potentially useful spin-off from the nuclear weapons program. The AEC hoped that nuclear explosions could, for example, be used to excavate a second Panama Canal.[8]

Article V of the Non-Proliferation Treaty said that nuclear weapon states would make the benefits of peaceful nuclear uses available to non-nuclear weapon states. PNEs for a time were seen as a potential nuclear offering to the non-nuclear weapon states in exchange for the latter's willingness to forego the option of acquiring nuclear weapons under the non-proliferation regime.[9]

India did participate in the development of the NPT although in the end it did not sign it when it was completed in 1968. This was in part because India detested the double standard built into the NPT, "nuclear apartheid" as some Indian political leaders called it. In an analogy to a drunkard emperor proscribing drinking in his empire, V. C. Trivedi, the Indian representative in Geneva in 1965, argued that the nuclear weapon states should concentrate on eliminating their own arsenals first before dictating to others. Trivedi proposed a two-stage treaty whereby, first, the nuclear weapon powers would agree not to transfer weapons or related technologies to others and would also cease all production of nuclear weapons and delivery vehicles, followed by reductions in these capabilities. Only then, in the second stage, would the non-nuclear weapon states agree not to acquire or build nuclear weapons.[10]

There were four major issues of concern for India: ending further production of nuclear weapons and delivery systems; securing commitments to pursue nuclear disarmament, while obtaining security guarantees; and retaining the right to conduct peaceful nuclear explosions. For their part, the nuclear weapon states offered little compromise. They concentrated their efforts on three narrower objectives: precluding the transfer of nuclear weapons and sensitive materials and know-how to other states; prohibiting receipt or production of nuclear weapons or explosive devices by previously non-nuclear weapon states; and making sure that non-nuclear weapon states accept safeguards on their activities to verify so that no diversion for weapons purposes would occur.[11] These three objectives were embodied, respectively in Articles I, II, and III of the NPT. In support of the Indian position on the question, Article VI of the treaty calls for nuclear weapon states "to pursue negotiations in good faith on effective measures relating to the cessation of the nuclear arms race at an early date and to nuclear disarmament."

In April 1964, India and Canada concluded an agreement for Canadian financing of the heavy-water-moderated CANDU (Canadian deuterium-uranium) power reactor to be built in the western state of Rajasthan by Indian engineers using Canadian technology and design assistance provided by Atomic Energy Canada, Ltd. This agreement provided evidence of India's dependence on Canadian know-how and capital at this stage of its nuclear program. Canada would provide half the uranium fuel elements for the initial load, and subsequent fuel as requested, in case India could not meet its goals for indigenous fuel production. The agreement also specified information sharing and assistance from Canada including blueprints. Canada would provide a loan of $37 million to cover foreign exchange expenses for Canadian technology and services.

As with the Indian-American negotiations over safeguards for the Tarapur plant, India and Canada engaged in difficult negotiations over the Rajastan nuclear project. Yet in this case a quicker solution was found. Canada accepted reciprocal Indian rights to inspect its Douglas Point (Ottawa) power station. India thought this arrangement equitable and agreed that the technology and fuel provided by Canada would be used only for peaceful purposes.

India has been repeatedly charged with having had designs to use the Canada-India reactor for military purposes, in violation of an undertaken given to Canada. In April 1965, the Canadian Minister for Industries, C. M. Drury, during a visit to Pakistan declared that Canada had secured a formal undertaking from India that the nuclear power reactor set up with Canadian assistance would be used only for peaceful purposes.[12]

Prestige and international respect were important motives behind Indian nuclear development. Acquiring a deterrent against both China and Pakistan was an important purpose as well especially since China had allied itself with Pakistan and supported it with bomb design and testing facilities. Only if the original five nuclear weapon states committed themselves to time bound negotiations for nuclear elimination was India prepared to abandon nuclear weapons.[13]

In 1974 India tested its first "peaceful nuclear explosive," ostensibly for civil engineering projects but really to create a deterrent against Chinese nuclear weapons. The two countries had fought a brief border war in 1962. Pakistan which had fought and lost three wars with India in the previous quarter century, felt obliged to match Indian nuclear weapons. Then Pakistan's Foreign Minister Zulfikar Ali Bhutto told the Guardian Commonwealth correspondent Patrick Keatley in March 1965 that his government was bound to work on the assumption that India could and might go nuclear without advance warning. Asked what course Pakistan would follow, Bhutto is quoted as having said: "Then we should have to eat grass and get one or buy one of our own."[14] From that time on the effort was focused on restoring the balance between the two countries and Pakistan therefore began a secret nuclear weapons program.[15]

In India, security concerns significantly affect the debate over arms control. India requires Pakistani reciprocity in any formal steps to reverse or even constrain its nuclear weapons capabilities. Similarly, India will require major improvements in its relationship with China, if not Chinese nuclear disarmament, before India will formally limit or roll back its nuclear weapons and ballistic missile capabilities.

Domestic sources of India's resistance to disarmament can be described under four categories. The symbolic importance of nuclear weapons and missile capability is the first reason for resistance to disarmament. By detonating nuclear bombs in 1974 and 1998, India demonstrated that it could do what only the five recognized major nuclear powers had done. The Indian achievement was even greater as it came against such odds and obstacles. The Indian press, television, and public have continued to glorify the achievements of the nuclear, and ballistic missile establishments. To constrain or surrender symbolically Indian weapons capabilities would go back on one of India's greatest perceived postcolonial achievements. Substitute symbols of scientific prowess would have to be created and promoted by India's leaders.

The second source of resistance has to do with colonialism. Indians equate the nonproliferation regime with colonialism and racism. So long as resistance to disarmament is identified with defying colonialism, it will be difficult to overcome.

The third source of grave reluctance to disarm is that nuclear weapons capabilities are political realities. Nuclear policy, especially since 1974, has become a potentially explosive political device deterring leaders from undertaking measures to restrain and roll back nuclear capabilities. India's political leaders fear that to abandon nuclear capabilities would be political suicide. They have been afraid even to adopt arms control measures that would limit the growth of these capabilities without requiring their abandonment. Public opinion limits India's willingness to undertake disarmament.

The final source of resistance are the nuclear and missile establishments which have become states within the state. The powerful role of the strategic enclave in propelling the development of nuclear weapons capabilities and aspirations in India must be taken into account. The strategic enclave has resisted disarmament policies regardless of India's security environment. As long as this enclave's achievements stand for India's postcolonial modernity and prowess, and as long as its performance and expenditures remain protected against disarmament, it will be a major force of resistance. The absence of independent institutions of scientific and technical expertise that could check and balance the work of the strategic enclave has undermined Indian decision makers for many years.[16]

The risk that the Indian strategic enclave is taking in preventing nuclear disarmament between India and Pakistan is great. Recent research finds significant consequences of any Indo-Pakistan nuclear war. Casualties from 50 KT weapon attacks on Indian and Pakistani cities would be high. In a combined attack there might be as many as 18 million casualties (37 percent of which would be fatalities) against Indian cities and 13 million casualties (42 percent of which would be fatalities) against Pakistani cities. The combined attack on all Indian strategic targets may result in 3.4 million casualties (31 percent of which would be fatalities) with 20 KT weapons and 6.7 million casualties (30 percent of which would be fatalities) with 50 KT weapons. Results against individual Pakistani military targets were similar to those of individual Indian targets. The combined attack on Indian transportation and power targets would result in 5.6 million casualties (34 percent of which would be fatalities) with 20 KT weapons and 10.6 million casualties (33 percent of which would be fatalities) with 50 KT weapons. The combined attack on all Pakistani transportation and power targets may result in 1.9 million casualties (40 percent of which would be fatalities) with 20 KT weapons and 3.1 million casualties (40 percent fatalities) with 50 KT weapons.[17] These numbers would be expected to rise as the current number of nuclear weapons—around a hundred thirty for India and probably one hundred for Pakistan—grows.

India continues to modernize its nuclear arsenal, with at least five new weapon systems now under development to complement or replace existing nuclear-capable aircraft, land-based delivery systems and sea-based systems. India is estimated to have produced enough military plutonium for 150 to 200 nuclear warheads but has likely produced only 130 to 140 warheads. Nonetheless, additional plutonium will be required to produce warheads for missiles now under development, and India reportedly is building several new plutonium production facilities. India's nuclear strategy, which was traditionally focused on Pakistan, now appears to place increased emphasis on China. [18] In 2004, an Indian Ministry of Defense official was quoted as saying that India in the next five to seven years would have 300–400 fission and thermonuclear weapons distributed to air, sea and land forces.[19]

At this point, we need to consider Pakistan and China and their nuclear weapon programs. Pakistani spokespersons in the United Nations and elsewhere were persistent in creating doubts about India's nuclear intentions. Those who understand that the 1974 nuclear test was to develop a deterrent bomb assumed that China's ambition to build itself up as a nuclear power was motivated entirely, or at least chiefly, by her desire to be accepted as the unchallenged overlord in South-East Asia—a desire in the fulfillment of which, it is further assumed, India constituted the main roadblock.

Those who take such a restricted view of China's nuclear motivations seem to overlook the broader picture of China's foreign policy. These may be broadly summed up under these heads: prestige; security or self-preservation; and a desire to establish Chinese hegemony in Asia. To this we could add a fourth motivation: the desire to rival Soviet influence in the Communist movement, especially in the Afro-Asian areas.[20]

Have China's nuclear explosions helped her in achieving her first objective, namely prestige? China's nuclear explosions have undoubtedly given her Great Power status, and the other nuclear powers are treating China with the deference due to a Great Power. But it is the United States, not India, whom China has always regarded as her primary enemy. India occupies a comparatively minor place in China's overall strategic thinking. In a statement issued on October 16, 1964, after the detonation of its first nuclear device, the Chinese government declared that China "will never at any time and under any circumstances be the first to use nuclear weapons."[21]

For its part, India has long adhered to a nuclear no-first-use policy, even though the policy was weakened by India's 2003 declaration that it could potentially use nuclear weapons in response to chemical or biological attacks, which would therefore constitute nuclear first use, even if it were in retaliation.[22]

In 1982, Premier Den Xiaoping began the transfer of nuclear weapons technology to Pakistan and, in time, to other third world countries. These transfers included blueprints for the ultrasimple CHIC-4 design using highly enriched uranium first tested by China in 1966. A Pakistani derivative of CHIC-4 apparently was tested in China on May 26, 1990.[23]

The history of Pakistan's nuclear program is the history of Pakistan. Fascinated with the new nuclear science, the young nation's leaders launched a nuclear energy program in 1956 and consciously interwove nuclear developments into the broader narrative of Pakistani nationalism. Then impelled by the 1965 and 1971 India-Pakistan wars, and more urgently by India's first nuclear weapon test in 1974, Pakistani senior officials tapped into the country's pool of young nuclear scientists and engineers and molded them into a motivated cadre committed to building the "ultimate weapon."[24]

On May 28, 1998, Pakistan announced the test of five nuclear explosive devices in the Chagai Hills in the western province of Baluchistan, a mere seventeen days after neighboring India had shocked the world with its first nuclear tests since 1974. Pakistan's response came as a surprise to many observers. Some had expressed doubts that Pakistan had the capability to construct a nuclear explosive device. But even those who thought that Pakistan could indeed test a weapon were astonished by the speed of the Pakistani reaction. Many observers wondered how a poor country recovering from catastrophic wars and national dis-membership when East Pakistan became Bangladesh—not to mention a nation struggling with national identity crises—could devote its limited state resources to acquiring such potentially destructive technology.[25]

Pakistan's enduring rivalry and strategic competition with India turned bitter over subsequent decades after a series of wars and crises. The last major war in 1971 resulted in humiliating military defeat and dismemberment of Pakistan, which simply reinforced its belief that its enemies were determined to destroy the very existence of the new state. This perception united the nation state into a "never again" mind-set that found consolation in the acquisition of a nuclear capability.[26]

Despite the many decades of the nuclear age in which the dangers and blessings of nuclear energy have revealed themselves, atomic weapons are considered a life-line for states like Pakistan and Israel, "orphan states" in the international system which operate outside the United States and other nuclear umbrellas.[27]

The person who spearheaded the idea of nuclear Pakistan was Zulfiqar Ali Bhutto. In Pakistan's early history there was no consensus about the desirability or utility of nuclear weapons. Only a few individuals, most notably Bhutto himself, believed that acquiring them was critical for Pakistan. However, following the devastating loss of East Pakistan in 1971 and the Indian nuclear test in 1974, opinions favoring nuclear weapons, held only by a minority, became national consensus—the necessity of nuclear weapons became a mainstream belief. This belief eventually determined the discourse that evolved gradually—first into developing a nuclear weapons capability that took some twenty-five years, and later operationalizing it after being forced to demonstrate that capability.

In the Indian case, the shock of losing the 1962 war with China combined with the Chinese nuclear test at Lop Nor in 1964 eventually led to the Indian test in 1974.

Today in Pakistan there are three important strategic beliefs regarding nuclear weapons. First, nuclear weapons are the only guarantee of Pakistan's national survival in the face of both a hostile India that cannot be deterred with conventional armed forces and unreliable external allies that fail to deliver in extreme need. Second, Pakistan's nuclear program is unfairly singled out for condemnation and international opposition because of its Muslim population. This feeling of victimization is accentuated by a belief that India consistently "gets away with" violating global nonproliferation norms. Third, is the belief that India, Israel or the United States might use military force to stop Pakistan's nuclear program: nuclear necessity for survival, international discrimination against Pakistan, and danger of disarming attacks.[28]

A. Q. Khan is a key figure in the development of Pakistan's nuclear weapon program. He was the world's leading black-market dealer in nuclear technology, described by a former CIA Director George Tenet as at least as dangerous as Osama bin Laden. A hero in Pakistan and revered there as the "Father of the Bomb," Khan built a global clandestine network that sold the most closely guarded secrets to Iran, North Korea and Libya. Khan has created havoc on any attempt to restrain one of the greatest threats facing the world today, the spread of nuclear technology.[29]

Khan was responsible not only for developing the nuclear capability in his native Pakistan but also for building an unrivalled, clandestine procurement network that spanned the globe. Many states have longed for the power that they perceived nuclear weapons would offer them, but the technical challenges appeared to them insurmountable to all but a few nations. At that point Khan began looking for customers and shifting his business model from import to export.

For many people, and not only those in his native Pakistan, Khan was—and remains—a hero. Many developing countries perceive a profound dishonesty in a handful of countries denying to others the technology that they refuse to relinquish. Critics argue that the United States in particular has attempted to maintain a restrictive cartel on nuclear weapons. For the West, the spread of the bomb may be a nightmare, heightening global insecurity and making it more likely that the weapon will be used. But for any individual nation facing its own particular challenges the bomb may represent the notion, real or imaginary, of security. For Pakistanis, Khan delivered the security and prestige that his country so desperately desired, and he was honored for it accordingly.[30]

Most countries with nuclear weapons have engaged in atomic espionage. Most have also proliferated nuclear technology based on perceived strategic priorities—the United States to the United Kingdom, France to Israel, the Soviet Union to China, China to Pakistan.[31] Khan was different. Here was an individual willing to proliferate to any country that was ready to pay.[32] And for the first time a dangerous array of products was available entirely in the private sector, outside of state control, creating what Mohammed El-Baradi, the head of the nuclear watchdog the IAEA, calls a Wal-Mart of private sector proliferation.[33]

In 1936 Abdul Qadeer Khan was born in Bhopal, India, to a Pakistani family. When the British granted independence to the Indian subcontinent in 1947, they used the partition of Hindus in India from Muslims in the split territories of East and West Pakistan as the fig leaf for their withdrawal. As with any partition, minorities were left behind on both sides. Life was grim for the Muslim Khan family in Bhopal; at the age of eighteen the young Khan migrated to Karachi in West Pakistan on foot.

Once in Karachi, Khan attended the D. J. Sindh College of Science, graduating in 1960 with a degree in metallurgy. After a brief stint in local government, Khan decided to pursue graduate studies in Western Europe. He met and married a Dutch girl, spent four years at the university in Delft, Holland, and emerged with a master's degree in metallurgical engineering. In the process, he became fluent in the Dutch and German languages. In 1968, the young Khan family moved to Leuven, Belgium. Khan entered Catholic University to continue his studies in metallurgy, in French.

Through all of those years, Khan's life had been that of the innocuous student and family man, but in 1971 events at home brought a sense of urgency to his life. In the spring of that year, the political leaders of East Pakistan rebelled, declaring independence from West Pakistan and adopting the name Bangladesh for their side of the subcontinent. A bloody civil war ensued, with India intervening on the side of the rebels. The Indian government wished to dismember, if not destroy, what it considered to be a rebellious remnant of old India. The war was a disaster for West Pakistan. In December 1971, her army surrendered with the Indians having taken ninety-three thousand troops prisoner. The military regime in West Pakistan fell, Zulfikar Ali Bhutto came to power, and all loyal Pakistanis (including the thirty-five-year-old Khan) decided they had to "do something" about India.

At home, the new government of Ali Bhutto had already decided what to do: go nuclear. There had been rumors of Indian nuclear ambitions for some time.[34] In January 1972, President Bhutto called together seventy of his leading scientists for a discussion of this option. The meeting took place in the Punjab. His audience was enthusiastic; they promised results within five years. In reality, it took more than two decades for Pakistan to achieve nuclear weapon status and then only with Chinese help.

Meanwhile, in Belgium, A. Q. Khan was about to receive his Ph.D. in metallurgy from Catholic University. In June he did so, and then he moved to Holland to take a job with a subcontractor working for the Uranium Enrichment Corporation. URENCO was organized by the British, Dutch and German nuclear power industries to develop the technology needed to separate U-235 fuel from natural uranium compounds. Ultra-centrifuges, rotating at very high speeds, were the preferred route. The resulting URENCO technology was the best in the world. In 1972, Dr. A.Q. Khan began collecting (i.e., stealing) that know-how with meticulous care.

U-235 is crucial to understanding URENCO technology. When a nucleus of uranium 235 undergoes fission, it splits into two smaller atoms and, at the same time, releases neutrons and energy. Some of these neutrons are absorbed by other atoms of uranium 235. In turn, these atoms split apart, releasing more energy and more neutrons.[35]

On May 18, 1974, the hated Indians tested a nuclear device under the Rajasthani desert. That was the final and defining moment for Khan. During the month that followed, he wrote a letter to President Bhutto explaining the role of the centrifuge in producing fissionable material. Khan offered to help with any Pakistani nuclear weapon program. He did not know that one was already underway. Bhutto responded with interest through his embassy in The Hague.

During the fall of 1974, Khan spent sixteen days at the URENCO facility. His day job was to translate documents, but during his spare time he toured the plant, taking notes on the design and operation of the equipment. His observations were written in his native Urdu to disguise them from curious eyes. In December 1974, President Bhutto and engineer Khan met in Karachi while Khan and his family were home for the holidays. The covert Pakistan Atomic Energy Commission had already embarked on a plutonium based nuclear weapon plan, but it was in trouble. At their meeting, Bhutto decided to put Khan in charge of a parallel enriched uranium program.

In the spring of 1975, Khan returned to his post in Amsterdam, now fully committed to collecting information and parts while Pakistani authorities began to purchase components for a nuclear enrichment program. These purchases began in August 1975.[36] A co-worker became suspicious; Khan was moved to a less sensitive job. In December 1975, he again returned to Pakistan for the holidays, but this time he never returned to URENCO. In the spring of 1976, Khan started to work at the Pakistan Atomic Energy Commission headed by Munir Ahmad Khan (no relation). Friction quickly developed. With the full support of the Bhutto government, Khan was authorized to organize the Engineering Research Laboratory in the town of Kahuta, fifteen miles due east of the Islamabad airport. That facility was to develop a uranium-enrichment capability for Pakistan. It opened for business on July 31, 1976.[37] Khan stayed in touch with his friends and informants in Holland, and his work may have received additional funding from Libya and Saudi Arabia. During the years that followed, China began the transfer of nuclear weapons technology to Pakistan—presumably to A. Q. Khan—in part as a consequence of the Chinese-Indian border clashes of the previous decade.

In the late 1970s, the American intelligence services learned of all this activity and CIA surveillance of A. Q. Khan apparently began in earnest. However, in December 1979, the Soviets invaded Afghanistan, and the Americans needed all the help they could get from neighboring Pakistan, so hard questions about covert nuclear programs were off the table. By the end of that decade, Khan had an operational enrichment centrifuge up and running at Kahuta and the Americans busied themselves elsewhere.[38]

Pakistan's nuclear arsenal is the fastest growing one in the world and raises concerns on many grounds. Although far from the scale of the Cold

War, South Asia is experiencing a strategic arms race. And the more weapons there are, the more potential for theft, sabotage and nuclear terrorism there is. Worries that Pakistani nuclear weapons technology might again be transferred to nuclear aspirants have not been laid to rest. Being outside the nuclear club makes it harder to ensure nuclear safety. Of gravest concern is the potential for a nuclear war, triggered by another large-scale terrorist attack in India with Pakistani state fingerprints on it, as in the 2008 Mumbai atrocity. This time such an event might be followed by an Indian army reprisal. Lowering the nuclear threshold, Pakistan has vowed to deter this with newly introduced battlefield nuclear weapons.[39]

Pakistan is currently enhancing its nuclear weapon capability across the board by developing and deploying new nuclear capable missiles and expanding its capacity to produce fissile materials for use in weapons.

Pakistan's estimated arsenal, which includes 140 to 150 warheads, could increase to a stockpile of 220 to 250 warheads if the current trend continues. A new nuclear-capable ballistic missile is being readied for deployment, and two nuclear-capable cruise missiles are under development. Two new plutonium reactors and a second chemical separation facility also are under construction. Although Pakistan's arsenal is clearly increasing several factors suggest that it may not have reached 140 warheads quite yet, but this stockpile will likely increase over the next ten years. Pakistan's enrichment capacity and stockpile are state secrets. Production capacity is estimated to be approximately 100 kg of weapons grade (90 percent) of Highly Enriched Uranium each year.[40]

The United States fears that the development of a full spectrum deterrent that is designed not only to respond to nuclear attacks but also to counter an Indian conventional incursion onto Pakistani territory, lowers the threshold for nuclear use in a military conflict with India.

In the Worldwide Threat Assessment for 2018, US Director of National Intelligence, Daniel R. Coates, said: "Pakistan continues to produce nuclear weapons and to develop new types of nuclear weapons, including short-range tactical weapons, sea-based cruise missiles, air launched cruise missiles, and longer-range ballistic missiles. These new types of nuclear weapons will introduce new risks for escalation dynamics and security threats in the region."[41]

The Trump administration assessment in 2018 was: "We are particularly concerned by the development of tactical nuclear weapons that are designed for use in the battlefield. We believe that these systems are more susceptible to terrorist theft and increase the likelihood of nuclear exchange in the region."[42]

Pakistan's weapons have been estimated to have yields of between 5 and 10 kilotons judging by its few nuclear tests. Pakistan claimed it conducted six tests on May 18 and 30, 1998, yet most experts have concluded that there were only two tests.[43]

Rejecting notions of "no first use," Pakistan reserves the right to use nuclear weapons against conventional attack. This is the basic premise of Pakistan's nuclear posture. Facing a potential enemy at whose hands it has three times suffered defeat and whose conventional superiority grows ever greater, Pakistan sees nuclear weapons as an equalizer. Pakistani officials also do not believe India's declared no first use doctrine. They assume that India would make use of nuclear weapons if it judged vital national interests were at stake.

Pakistan does say that it will not "use or threaten to use" nuclear weapons against any non-nuclear weapon state unless that state joins a hostile military coalition with nuclear armed states.[44]

Finally, it is important to present the risk to the climate and environment of the world if a full-scale nuclear conflict were to break out between India and Pakistan. Recent research finds significant nuclear winter effects from an India-Pakistan nuclear war involving 100 weapons (50 per side) of 15 kt yield dropped on major cities.[45]

The central point of the new findings is that the long-term consequences of nuclear war could constitute a global climatic catastrophe. The Hiroshima bomb that killed between 100,000 and 200,000 was a fission device with the explosive equivalent of 12,000 tons of TNT. Many weapons with yields roughly equivalent of the Hiroshima bomb are today assigned to "tactical" or "theater" military missions.

It is generally accepted that a "small" nuclear war, such as one between India and Pakistan, would be almost impossible to contain before it escalated to include China's arsenal as well as others.[46]

Recent estimates of the immediate deaths from blast, radiation, and fires in a major exchange in which cities were targeted range from several million to 1.1 billion people. Serious injuries requiring immediate medical attention (which would be largely unavailable) would be suffered by a comparably larger number, perhaps an additional 1.1 billion. Thus, it is possible that something approaching half the population of the planet would be killed or seriously injured by the direct effects of the nuclear war.

Then there would be the social disruption. Electricity would be unavailable as well as fuel, transportation, food deliveries, communications and other civil services. Medical care would be absent and as a result of the decline in sanitation measures, rampant diseases and severe psychiatric disorders would probably lay claim to a significant number of further victims.

The destruction of cities in South Asia such as Delhi or Islamabad can be accomplished, as was shown at Hiroshima and Nagasaki, by lower yield explosions less than a kilometer above the surface. Low-yield airbursts over cities or near forests will tend to produce massive fires, some of them over 100,000 square kilometers or more. City fires generate enormous quantities of black oily smoke which rises at least into the upper part of the lower atmosphere, or troposphere. If firestorms occur, the smoke column rises vigorously, like the drafts in a fireplace, and may carry some of the soot into the lower part of the upper atmosphere, or stratosphere. The smoke from forest and grassland fires would initially be restricted to the lower troposphere.[47]

The fission of the (generally plutonium) trigger in every thermonuclear weapon and the reactions in the (generally uranium 238) casing added as a fission yield "booster" produce a witch's brew of radioactive products, which are also entrained in the cloud. Each such product, or radioisotope has a characteristic "half-life" (defined as the time to decay to half of the original level of activity). Most of the radioisotopes have very short half-lives and decay in hours to days. Particles injected into the stratosphere, mainly by high yield explosions, fall out very slowly—characteristically in about a year, by which time most of the fission products, even when concentrated, will have decayed to much safer levels. Nuclear explosions from weapons between 100 kilotons and one megaton generate a radiant fireball, which will partially extend into the stratosphere. The high temperature in the fireball chemically ignites some of the nitrogen in the air, producing oxides of nitrogen, which in turn chemically attack and destroy the gas ozone in the middle stratosphere. But ozone absorbs the biologically dangerous ultraviolet radiation from the sun. Thus, after the soot and dust have settled out the partial depletion of the stratospheric ozone layer by high yield nuclear explosions will increase the flux of solar ultraviolet radiation. An increase of the solar ultraviolet flux at the surface of the Earth is potentially dangerous for life.

These four effects: obscuring smoke in the troposphere, obscuring dust in the stratosphere, the fallout from radioactive debris, and the partial destruction of the ozone layer constitute the four known principal adverse environmental consequences that could occur after a nuclear war is "over." There may be others about which we are still ignorant. The dust and, especially, the dark soot absorb ordinary visible light from the Sun, heating the atmosphere and cooling the Earth's surface.[48]

There is now substantial evidence that the heating by sunlight of atmospheric dust and soot over northern mid-latitude targets would profoundly change the global circulation. Fine particles would be transported across the equator in weeks, bringing the cold and the dark to the Southern Hemisphere. While it would be less cold and less dark at the ground in the Southern Hemisphere than in the Northern, massive climatic and environmental disruptions may be triggered there as well.[49]

The explosion of the Tambora volcano in Indonesia in 1815 led to an average global temperature decline of only 1 degree Celsius, due to the obscuring of sunlight by the fine dust propelled into the stratosphere, yet the hard freezes in the following year were so severe that 1816 has been known in Europe and America as "the year without a summer." A 1-degree Celsius cooling would nearly eliminate wheat growing in Canada.

Even a comparatively small nuclear war, such as one between India and Pakistan, could have devastating climactic consequences, provided that cities are targeted.[50]

Since the atomic bombs fell on Hiroshima and Nagasaki, ending the Second World War, we have been spared the tragedy of nuclear war. Therefore, the potential effects of such a war must remain a matter of speculation. But we do know what has happened at accidents at nuclear reactors, and this will give us some idea of what a nuclear war would bring us. The next chapter, therefore, will be a study of the disasters at Mayak in the southern Ural Mountains in what was then the USSR and in the United States around Hanford Reservation. Both of these sites produced plutonium for use in atomic bombs and were therefore crucial to the arms race in the cold war.

NOTES

1. Baum, Seth D. 2019. "Why catastrophes can change the course of history," Global Catastrophic Risk Institute, http://www.bbc.com/future/story/20190408-how-catastrophes-can-change-the-path-of-humanity
2. Sagan, Carl. 1983. "Nuclear War and Climatic Catastrophe: Some Policy Implications," *Foreign Affairs,* Vol. 62, Iss. 2 (Winter 1983), p. 259; Toon, Owen, Robock, Alan, and Turco, Richard P. 2008. "Environmental consequences of nuclear wars," *Physics Today,* Vol 61 (12), December 2008, pp. 37-42; Mills, Michael J., Toon, Owen, Lee-Taylor, Julie, and Robock, Alan. 2014. "Multidecadal global cooling and unprecedented ozone loss following a regional nuclear conflict," *Earth's Futures* 100, Vol. 2, Issue 4, April 2014, pp. 161-176, https://doi.or/10.1002/2013EF000205.
3. Baum, Seth D. 2015. "Confronting the threat of nuclear winter," *Futures* Vol. 72, Sept. 2015, pp. 69-79; Mills, M. J., Toon, O. B., Lee-Taylor, J., and Robock, A. 2014. "Multi-decadal global cooling and unprecedented ozone loss following a regional nuclear conflict," *Earth's Future* 2 (4), pp. 161-176.
4. Baum, "Why catastrophes can change the course of history," Op. cit.
5. Toon, O. B., Turco, R. P., Robock, A., Bardeen, A. C., Oman, L., and Stenchikov, G. L. 2007. "Atmospheric effects and societal consequences of regional-scale nuclear conflicts and acts of individual nuclear terrorism," *Atmos. Chem. Phys.* 7, 1973-2002; Robock, A., Oman, L., and Stenchikov, G. L. 2007. "Nuclear winter revisited with a modern climate model and current nuclear arsenals: Still catastrophic consequences," *Journal of Geophysical Research: Atmospheres,* Vol. 112, No. D. 13, July 6, 2007, 1-14.
6. Appadorai, A. and Rajan, M. S. 1985. *India's Foreign Policy and Relations,* New Delhi: South Asian Private, p. 80.
7. Perkovich, George. 1999. *India's Nuclear Bomb. The Impact on Global Proliferation.* Berkeley: University of California Press, p. 479.
8. Perkovich, p. 90.
9. Tannenwald, Nina. 2007. *The Nuclear Taboo. The United States and the Non-Use of Nuclear Weapons since 1945.* Cambridge: University Press, p. 269.
10. Statement by Indian representative (Trivedi) to the Eighteen Nation Disarmament Committee, August 12, 1965, in US ACDA Documents on Disarmament, pp. 335-336.
11. Shaker, Mohamed J. 1980. *The Nuclear Non-Proliferation Treaty: Origin and Implementation,* 1959-1979, 3 vols. New York: Oceana.
12. Mirchandani, G.G. 1968. *India's Nuclear Dilemma,* New Delhi: Humanities Press, p. 179.

13. Rhodes, Richard. 2010. *Twilight of the Bombs. Recent Challenges, New Dangers and the Prospects for a World without Nuclear Weapons*, New York: Knopf, p. 230.
14. Mirchandani, G. G., p. 179.
15. Dyer, Gwynne 2016. *War The New Edition* Toronto: Vintage Canada, pp. 427-8.
16. Perkovich, pp. 457-458.
17. Batcher, Robert. 2004. "The Consequences of an Indo-Pakistani Nuclear War," *International Studies Review* Vol. 6, No. 4, Dec. 2004, p. 161.
18. Kristensen, Hans M. and Kordu, Matt. 2018. "Indian nuclear forces, 2018," *Bulletin of the Atomic Scientists*, Vol. 74, issue 6.
19. Vivek, Raghuvanshi. 2004. "India to Stay the Course on Nuke Doctrine," *Defence News*, November 1, 2004.
20. Ibid, p. 61.
21. Ibid, pp. 64-68, Reed, Thomas C. 2008. "The Chinese Nuclear Tests 1964-96," *Physics Today*, Vol. 61 (9) Sept. 2008, 47-53.
22. Sundaram, K. and Raman, M. V. 2018. "India and the Policy of No First Use of Nuclear Weapons," *Journal for Peace and Nuclear Disarmament* 1 (1), 152-168.
23. Smith, R. Jeffrey and Warrick, Toby, 2009. "Pakistan nuclear scientists' accounts tell of Chinese proliferation." www//washingtonpost.com/wp-dyn/content/article/2009/11/12/AR 2009 111211060.html.
24. Khan, Feroze H. 2012. *Eating Grass: The Making of the Pakistani Bomb.* Stanford, CA: Stanford University Press, abstract.
25. Ibid, p. 1.
26. Ibid, p. 2.
27. Mandelbaum, Michael. 1995. "Lessons of the Next Nuclear War," *Foreign Affairs* 74 (2), March-April 1995, 22-35; Corera, Gordon. 2006. *Shopping for Bombs: Nuclear Proliferation, Global Insecurity, and the Rise and Fall of the A. Q. Khan Network.* New York: Oxford University Press.
28. Khan, p. 6.
29. Corera, p. xiii.
30. Ibid, p. xiv.
31. Richelson, Jeffrey T. 2006. *Spying on the Bomb: American Nuclear Intelligence from Nazi Germany to Iran and North Korea*, New York: Norton.
32. Jehl, Douglas, 2004. "CIA Says Pakistan Gave Iran Nuclear Aid," *The New York Times*, November 24, 2004.
33. Landler, Mark and Sanger, David E. 2004. "Pakistan Chief Says It Appears Scientists Sold Nuclear Data," *The New York Times*, January 24.
34. Reed, Thomas, C., and Stillman, Danny B. 2009. *The Nuclear Express*, Minneapolis: Zenith pp. 152-153.
35. Bulgac, Aurel; Jin, Shi; Stetcu, Ionel. 2020. "Nuclear Fission Dynamics: Past Present, Needs, and Future," *Frontiers in Physics* 8: 63, doi: 10.3381/phy. 2020.00063.
36. Ur-Rehman, Sahid. 1999. *The Long Road to Chagai.* Islamabad: Printwise Publications, 1999.
37. Reed and Stillman, p. 154.
38. Ibid.
39. International Institute for Strategic Studies. 2013. Adelphi Series. Chapter 1, "Pakistan's nuclear programme," 53:443,13-46 DOI: 10.1080 19445571.2013.901482.
40. Mian, Zia et al. 2006. "Fissile Materials in South Asia; The Implications of the US-India Nuclear Deal," research report of the International Panel on Fissile Materials (IPFM), September 2006, p.15.
41. Kristensen, Hans M. et al., Pakistan nuclear forces, 2018, *Bulletin of the Atomic Scientists*, Vol. 74, Issue 5, 2018. Special Issue: the verification of arms control agreements.
42. Siddique, F. and Faisal, M. 2016. "Pakistan's Strategic Nuclear Policy and Implications for Deterrence Stability," *CISS Insight: Quarterly News and Views* IV (1): 1-17; *Economic Times* 2017, "US Worried Pakistan's Nuclear weapons could land up in Terrorists' hands," August 25, 2017, https//economictimes.indiatimes.com/new/defence/

US-worries-pakistans-nucIbid,learweapons-could-land-up-in-terrorists-hands-official/articleshow/60220358.cms.

43. Kristensen, 2018.

44. Lavoy, Peter, Islamabad's Nuclear Posture: Its Premises and Implementation, in: Sokolski, Henry D. S. (ed.). 2008. *Pakistan's Nuclear Future: Worries Beyond War,* Carlisle, PA: Strategic Studies Institute, p. 158. Geller, Daniel S. 2003. "Nuclear Weapons and the Indo-Pakistani Conflict: Global Implications of a Regional Power Cycle." *International Political Science Review*. Vol. 24, No. 1, Jan. 2003.

45. Mills, M. J., Toon, O. B., et al., pp. 161-176.

46. Sagan, Carl, 1983. Nuclear War and Climatic Catastrophe: Some Policy Implications, *Foreign Affairs* Vol. 62, 2, (Winter 1983), 251.

47. Ibid, 263.

48. Ibid, 264.

49. Ibid, 265.

50. Ibid 267.

FIVE

The Environmental and Human Cost of Producing Nuclear Weapons

The nuclear arms race between the United States and the Soviet Union meant that for both sides there was a great need to produce nuclear warheads. For both the Americans and the Soviets the process of creating these weapons often needed plutonium. The two most important sources for this plutonium were plants at Hanford for the United States and at Mayak for the Soviets. Serious problems including pollution and accidental release of radiation plagued both the Hanford plutonium plant in Washington State and the Mayak plant in the Urals. It is to this sad history that we will now focus our attention.

Plutonium does not occur (in useful amounts) anywhere in nature, at least anywhere that has been discovered yet. For a nuclear weapon to work an explosive chain reaction must take place.[1] Of all the steps on the nuclear weapons assembly line plutonium production is the dirtiest. Each kilogram of the final product generates hundreds of thousands of gallons of radioactive waste. Despite the many problems, due to the exigencies of the Cold War, plant engineers were pushed to produce as much plutonium as possible as quickly as possible, and they polluted the surrounding landscape freely, liberally and disastrously, all the while putting their workers and the surrounding population at risk of nuclear contamination.[2]

In four decades of operation, the Hanford plutonium plant near Richland and the Mayak plant next to Ozersk each emitted at least 200 million curies of radioactivity—twice what issued from Chernobyl—into the surrounding environment. A curie (Ci) is a quantity of radioactivity, originally defined by the number of radioactive disintegrations in one gram of radium (approximately 37 billion disintegrations per second). The Ci has been replaced by a newer standardized unit, the Becquerel.[3] Estimates at the Mayak plant are far greater, at 1 billion curies.[4]

The plants left behind hundreds of square miles of uninhabitable territory, contaminated rivers, ruined fields, and thousands of people claiming to be sick from the plants' radioactive emissions.

Before Chernobyl and Fukushima Daiichi came Hanford and Mayak. Chernobyl and Fukushima are household names. Why have so few people ever heard of Hanford and Mayak?[5]

The political hostilities that fueled the Cold War have ended, but the nuclear chapter of world history is far from over. The lethal landscape surrounding the plutonium plants is pockmarked with landmines of percolating radioactive isotopes that self-heat to hundreds of degrees, corrode metals, and seep easily through soil to be taken up by plant life—and will do so for tens of thousands of years.[6]

Nestled in a crook of the Columbia River in a dry, almost empty corner of south-central Washington State, the 570 square mile reservation known as the Hanford Reach was the site of an "atomic city" chosen after a nationwide search by the US Army for the top-secret Manhattan Project. There, beginning in 1943, 45,000 workers built the three production reactors that isolated the plutonium for the July 1945 test explosion near Alamogordo, New Mexico, and for the bomb dropped on Nagasaki the next month. Crucial to the choice of this site was huge quantities of cold water and electricity. The tiny riverside village of Hanford was settled on by the bomb makers because the surroundings offered access to the coolest big river in the United States and power from the nation's biggest dam.[7] An even more important reason for this choice in the words of Franklin T. Matthias, the Army lieutenant colonel who scouted locations for a plutonium plant in 1942, was that Hanford was "an area with almost no people."[8] There would be few victims—or witnesses—in case the biggest secret of World War II happened to blow up.[9]

The army ordered the few hundred farmers and fruit growers who lived in and around Hanford to get out within thirty days. It bought up about six hundred square miles of what it considered to be wasteland and sealed it off behind barbed wire and machine gun armed men.

In the hectic months of 1945 alone, 340,000 curies of radioiodine (I-131), generated by the chemical separation of plutonium from the irradiated fuel rods, were released into the atmosphere. This gas blew down the wind streams ("downwind") of eastern Washington State: to the east, northeast, and southeast. The radioiodine also deposited itself on vegetation throughout the region where it was available to cows and goats, which produced milk drunk by humans. Radioiodine, if inhaled, ingested, or absorbed through the skin, affects the thyroid, which is particularly radiosensitive in children. As a carrier of radiation to the thyroid, the milk pathway is many times more potent than simple inhalation of I-131.[10] According to the newly declassified Hanford documents released in 1986, among the most serious problems were the unfiltered airborne releases of iodine 131 from the reprocessing operations which totaled 537,000 curies from 1944 to 1955, with 340,000 curies emitted in 1945 alone.[11] Heavy iodine releases from 1943 to 1946 were attributed to the war emergency, when the cooling time was too short for most of the iodine 131 to decay. But iodine releases continued to be a chronic problem until 1956.[12] Until recent years, area residents were not informed of the discharges nor warned of any potential dangers, even when releases far exceeded the tolerable limits and "maximum possible concentrations" (MPCS) defined as safe at the time. In fact, Hanford scientists and managers, on numerous occasions throughout the first four decades, specifically told the public that the plant's workings and wastes were well controlled and harmless.[13]

Hanford was transformed and for the next twenty-five years after Nagasaki happened, great complexes of production reactors and chemical plants turned out tens of thousands of kilograms of plutonium for the USA's swollen stockpiles of nuclear weapons. In the process, the chemical plants also turned out more than 70 million gallons of intensely radioactive waste into solid cakes of salt and stored the cakes in steel tanks; 42 million gallons of the waste are still in liquid form. All of this radioactive detritus will have to be guarded carefully for centuries until it decays and becomes harmless.[14]

There are three principal stages in the production of a nuclear weapon. In the first stage, blocks of uranium are irradiated (bombarded by neutrons) in the reactor to produce plutonium. This process produces other, less useful, radioactive isotopes, which have to be separated from the uranium and plutonium in the second stage of production. Separation takes place in the radiochemical plant, where the various radioactive substances are dissolved in nitric acid, chemically treated, and filtered in successive cells. In the third stage of the process, which takes place in the metallurgical plant, the separated plutonium is converted into the particular metallic forms required by nuclear weapons.[15]

For twenty-five years there were nine reactors in operation at the Hanford plant, which produced several tens of thousands of kilograms of plutonium.[16] The recovery of plutonium was the main purpose of the entire production process, and the cost of producing one kilogram of plutonium is twenty to thirty thousand dollars in the United States.[17] Despite the value and importance of plutonium the techniques of plutonium separation have not been perfected to the point of 100 percent efficiency.

It was at the Hanford nuclear complex for the production of plutonium that two very dangerous situations developed.

The first accident is known as the Tank 106-T leak in which 435,000 liters of concentrated liquid waste escaped. In June 1973 it was noticed that one of the larger underground tanks, tank 106-T, had developed a leak. How long the concentrated radioactive liquid had been escaping remains unknown, nor do we know how much of its caustic, boiling contents had seeped into the sandy soil near the center of the reservation. As a matter of fact, no one was certain how much liquid had been in the tank in the first place. The liquid had penetrated the soil and had already spread contamination quite widely and to a depth of over forty feet below the tank. Fortunately, the radioactivity did not penetrate the water table two hundred feet below and therefore was not carried to the nearby river. After discovering the leak the liquid remaining was pumped out, but an estimated 115,000 gallons had run off into the soil. The leaked waste contained 40,000 curies of cesium-137, 14,000 curies of strontium-90 and 4 curies of plutonium, as well as a smaller quantity of other isotopes. Cesium 137, or radiocesium, is formed as one of the more common consequences of uranium 135 in nuclear reactors and nuclear bombs.[18]

It was only around noon on Saturday, June 9, 1973, that federal authorities and Atlantic Richfield Hanford Company, which was in charge of the day-to-day management of the nuclear garbage, began to grasp the magnitude of the problem. Technicians calculated that the seepage had begun "on or about" April 20, 1973. For fifty-one days thereafter, roughly 2,500 gallons of liquid waste had dribbled out of the steel-and-concrete tank each day.[19] The investigation of the incident revealed that other tanks also had leaks and that over a long period (since 1958) no less than half a million curies had leaked into the soil under the tanks. In this zone there were 151 containers for the liquid waste, whose total volume in 1973 came to tens of millions of curies.

If such a leak had occurred in an area where the water table was closer to the surface, the ground water would have quickly spread the contamination to a vast area, and after a certain time it would have worked its way into the soil, the vegetation and the lakes and ponds of the surrounding region. Dozens of kilometers of territory with such a high contamination level could then have become the source of extensive secondary contamination through soil erosion, dust storms and other causes.

The amount of plutonium lost was rather small as reckoned in curies. But with a half-life of 240,000 years, even four curies of the manmade element is a serious quantity. But where did the plutonium come from? The radioactive mixtures are supposed to be "waste" remaining after the separation of plutonium.

According to an official report, the WASH 1520 Report of the United States Atomic Energy Commission, which investigated the problem and recommended the removal of plutonium-contaminated soil with special equipment, conditions could have affected this high concentration of plutonium in such a way as to trigger a chain reaction, resulting in an explosion. A chain reaction could have been set off if water soaked into the plutonium rich soil. The rapid heating of the water could turn it to steam and the pressure of the steam could produce an explosion, discharging the radioactive soil to the surface. Nevertheless, it was not enough to cause an explosion, especially because the plutonium was dispersed through a large amount of soil .[20]

Other high-level waste tanks have also leaked. Between August 1958 and June 1973 an estimated 422,000 gallons containing more than half a million curies seeped out of 15 other tanks, all of which have since been "retired." But the leak in 106-T was something different. It was the largest single accidental release of radioactive waste in the AEC's history up to that time.

What the incident reveals about the oversight of the AEC over the US' ever-expanding store of nuclear processing wastes, 75 percent of which are stored at Hanford, is also significant. How could the AEC lose the equivalent of a railroad tank car full of radioactive liquid hot enough to boil itself for years on end and knock a Geiger counter off scale at a hundred paces?

A 129-page report on the causes of the June 1973 leak is available. The report, written by a four-man committee, attributes the accident to aging tanks and primitive monitoring technology, but mostly to managerial laxity and human error on the part of Atlantic Richland. The report contains a brief admission that the AEC's Richland operations office, which is supposed to supervise Atlantic Richland contractors, failed to detect flagrant deficiencies in the management of Hanford's thirteen waste storage tank "farms."

There are two ways of detecting a leaking tank: the level of fluid in a tank diminishes, or radioactive material is detected while leaking out. To deal with the first, tank farm operators were supposed to take weekly readings of fluid levels. To deal with the second, they were supposed to take weekly or monthly radiation readings at dry wells spotted around the tanks. If fluid levels sank and radiation in the wells rose, that meant that a tank was leaking.

The problem, according to the report, was that the operators did not know how to interpret the readings; and a day shift supervisor in charge of half of Hanford's tanks, who did know how to read the data, let 6 weeks-worth of charts and graphs pile up on his desk because of "the press of other duties." A "process control" technician elsewhere at Hanford, who was supposed to be reviewing the tank readings for "long-term trends," received no data for more than a month. This technician emerges as a "hero" in this sad story. Fragmentary readings of fluid levels in Tank 106-T arrived in his hands-on June 7, 1973, but it was enough to show him that something was amiss. The technician put out the alarm; the supervisor confirmed the leak the next morning after checking his records and promptly resigned.

The tanks were wearing out. 106-T was built in 1943–1944 and 108 others still in use are more than twenty years old. The AEC knew all this. In spite of all precautions more spills from Hanford's worn out tanks are inevitable. On July 6, 1973, yet another one sprang a leak of high-level waste. This time, tank farm crews were alert. They held the loss to 1,500 gallons.[21]

The second important incident at Hanford was a near-disaster at the Hanford reservation. This was not in the tanks with highly active liquid waste, which has just been described, but in one of the trenches where the less active waste was poured. The volume of this low-level waste was much greater, and it would have been very costly and required very large tanks to store the perhaps one billion liters of this liquid. Therefore, a simpler and cheaper means of storage was worked out. The waste was simply allowed to soak into the dry earth that constituted the floor of the trench. Since the water table was so deep, it was felt that the ground water would prove to be sufficiently stable to deal with all the radioactive liquids. However, the process taking place in the ground over a period of many years created an unexpected situation in one of the large trenches.[22]

Plutonium was present in the isotope mixture poured off in less concentrated forms into the trenches. It was assumed that this plutonium, since it is a poorly soluble element, would precipitate out, be largely absorbed by the soil beneath each trench, and become fixed in a nonsoluble state. And that is what happened. But it occurred in a much narrower layer of soil than expected. The absorption of the radioactive mixture by the soil took place along the lines of column chromatography (an ion exchange process). In 1943–1945, column chromatography was not yet known. Different substances were separated by the soil at different depths, depending on their properties and molecular weights. The plutonium was absorbed and accumulated in a relatively thin, upper layer of the soil rather than spread evenly through the whole mass. Over a period of many years, approximately 100 kilograms of plutonium accumulated not far below the dirt floor of one of these trenches (Z-9). This was a quantity sufficient to produce nearly a hundred small-size atom bombs each with a destructive force equal to the bombs dropped on Japan in 1945. The volume of soil containing plutonium at trench Z-9 was approximately 1,800 cubic feet.

The AEC estimated that the necessary mining equipment would cost about $2 million, and this sum was appropriated for the work on trench Z-9. It was proposed that the plutonium be separated from the soil, since 100 kilograms had a commercial value of over $3 million. But this was never done, and the soil dug from trench Z-9 was simply stored again elsewhere.

Altogether there were eleven trenches of this type at Hanford, no critical situation developed at the ten other trenches because, according to the experts' estimates, those trenches altogether contained only 200 kilograms of plutonium.

From the time this atomic complex was founded, several million gallons of liquid waste had been poured into these trenches each year.[23] Some 440,000 cubic yards of high-level and transuranic wastes—about two-thirds of all the waste generated from warhead production in the United States—is stored above the water table at Hanford.

The most startling revelation disclosed in the newly declassified Hanford documents was of a December 1949 experiment that deliberately contaminated eastern Washington State. This was called the "green run" by General Electric Nucleonics Department for the Atomic Energy Commission.

The "green run" reprocessed during a twenty-four-hour period one ton of irradiated uranium that had cooled only sixteen days. Uranium was normally cooled for 83 to 101 days before reprocessing began.

The green run released into the atmosphere some 5,500 curies of iodine 131 secretly measured by the Atomic Energy Commission in a two hundred-by-forty-mile plume from Spokane Washington to the Dalles, Oregon. By way of contrast, when the 1979 Three Mile Island accident released 15 to 75 curies of radioactive iodine into the Pennsylvania countryside, people around Harrisburg were evacuated and milk was impounded.[24]

The reason for the "green run" was to aid military efforts to detect plutonium production in the Soviet Union. Carl Gamertsfelder, an American physicist who worked in the field of health physics during the Manhattan Project, told the *Oregonian* newspaper that the experiment was part of an elaborate Cold War plan to locate plutonium plants inside the Soviet Union. He said that the AEC assumed the Soviets were rushing to produce nuclear bombs and were thus using short cooled fuel—about twenty days out of the reactor—which might be easily detected at considerable distances from Soviet separation plants. They decided to test their assumptions in eastern Washington by creating radioactivity levels in the Hanford area similar to those that presumably existed near Soviet plants.

Gamertsfelder said the experiment was carried out by the Army and the AEC despite Hanford officials' concerns that poor weather would unnecessarily expose offsite populations. The Hanford Historical documents also show that the green run was merely part of a much larger pattern of contamination. From 1947 on, monthly airborne releases of iodine 131 averaged from 100 to 2,000 curies. In 1951, failure of the silver reactors—a chemical reactant to remove iodine—in a reprocessing plant resulted in releases totaling 19,000 curies. In the 1950s, each operating reactor was releasing up to 1,600 curies of beta products per day to the river through its cooling water, with trace activity documented as far away as Portland, Oregon. The fission product zinc 65 was detected in oysters three hundred miles downstream in Wallapa Bay.

The REDOX (reduction oxidation) plant which operated from 1952 to 1967 began to be a problem as early as 1952, with at least nine major releases, mostly of ruthenium 106-contaminated ammonium nitrate flakes, occurring over two years. The emissions were so heavy on the nuclear reservation, reaching sixteen rems per hour, that Hanford managers restricted access to protect workers. Doses in other parts of the nuclear reservation were measured as high as five rems per hour-today's maximum annual dose for a radiation worker. A rem is a short form for Roentgen equivalent man and qualifies the health effects of exposure to ionizing radiation. The rem measures dose equivalency and is calculated using different factors, including the absorbed dose and the type of radiation involved. The rem has been replaced by the standard international unit, the Sievert (Sv).[25]

Hanford rejected a proposal to issue a health warning to nearby towns. Because Ruthenium 106 is a beta emitter with a half-life of one year, health physicists were worried that ruthenium flakes could cause tissue damage if they came in direct contact with human skin, or increased cancer risks if inhaled or ingested.[26] A beta particle, also called beta ray or beta radiation, is a high energy, high speed electron emitted by the radioactive decay of an atomic nucleus during the process of beta decay.[27]

The Centers for Disease Control and Prevention (CDC) reviewed 19,000 pages documenting regional radioactivity from Hanford and concluded that the releases were the largest documented from any US nuclear facility to date.[28] In September 1986, the CDC convened the Hanford Effects Panel in Richland, together with Washington State radiation officials and area Indian tribes. The CDC found preliminary evidence that radiation doses to the thyroid from iodine 131 for people living near Hanford in the 1940s, were large enough to cause health problems, particularly in infants. The exposures may have ranged from 50 rems annually to the thyroid gland in adults in Spokane to 2,300 rems in infants on the perimeter of the nuclear reservation. The Environmental Protection Agency guidelines today call for withholding milk and produce when thyroid doses of from 1.5 to 15 rems are anticipated.[29]

The task of reconstructing the effects of the early Hanford emissions will be difficult because Hanford officials did no studies of offsite populations, and no tumor registry was ever established in eastern Washington.[30] The secretary of energy formally acknowledged in 1990 that Hanford's releases of iodine-131 were high enough to cause illness among people living downwind. Seven years of federal research on the size and spread of the releases have found that about eighty thousand people were exposed to more than 10 rads of radiation. A rad is a measure of absorbed radiation. 10 rads would be roughly equivalent to a dozen chest X rays. The current limit in the US for an annual safe dose of man-made radiation is 0.025 rads. These guidelines call for evacuation if the dose to the thyroid reaches 5-25 rads.

As for workers at Hanford, they have a long history of being misled about the risk of radiation in their workplace. The AEC knew in 1947 that Hanford workers were exposed to "significant quantities" of radioactive particles. But it chose not to inform them even as it secretly characterized the exposure as a "very serious health problem."[31]

Today the PUREX reprocessing plant and the associated finishing plant, also known as Z-plant, which converts plutonium nitrate solution from PUREX into metallic "buttons" for shipment offsite, are the only operating weapons facilities on the reservation at the present time.

The cost of cleaning up Hanford may prove to be the most expensive such project in history. The 2019 Hanford Lifecycle Scope Schedule and Cost report estimated the remaining cost of environmental remediation of the former plutonium production complex in Washington State at $323.3 billion under a best-case scenario. The worst-case would be $677 billion.[32] It remains a serious environmental question whether such a large expenditure will ever be made.

The most difficult dilemmas at Hanford began after the manufacturing plant achieved its initial success. In February 1986, the Department of Energy, the successor agency to the Atomic Energy Commission, released approximately nineteen thousand pages of reports pertaining to the early history of Hanford.[33] From these documents and from nearly fifty thousand additional pages of documentation released over the next six years the world learned of immense waste discharges from the Hanford facility. Radioactive and chemical wastes totaling into the billions of gallons of liquids, and billions of cubic meters of gases were emitted from the plant after its start-up in late 1944. While trying to contain communism, the Hanford site was spreading millions of curies of radioactivity into the Columbia River and into the air and soil of the Columbia Basin.

The Hanford site was for many decades the core of the US "atomic shield." Its achievements had been therefore immense. Yet now its waste volume is overwhelming. It has done so much good and yet it has also done so much damage.

The evidence that such wastes were discharged in large quantities without public knowledge or warning raises fundamental questions about American democracy. Chief among these questions is that of secrecy and national security. The secrecy surrounding the production of the first atomic bombs and the early years of the nuclear weapons build-up also until recently concealed the fact that radioactive byproducts of Hanford plutonium extensively contaminated portions of the Pacific Northwest. The emissions were heaviest during the war emergency, but continued for two decades after the war had ended—possibly doing long-term damage to people living downstream from the production reactors. The public was not warned about the offsite contamination, which was measured but kept secret in classified documents, most of which were declassified for the first time in 1986.[34]

At the Hanford site today, a renaissance is in progress. An extensive waste cleanup effort, pioneering innovative technologies, has begun. The open disclosure of Hanford's historical record has facilitated this clean up. The past now serves the present and makes possible positive change.[35]

The first plutonium settlement in the USSR was no Richland. There was no master plan, no scheme to segregate unreliable prisoners and soldiers from the upstanding, trained employees who were entrusted with state secrets. As the conscripted workers caroused and lived off the surrounding countryside, they spread word about the expanding "secret" installation going up in the forest.

The regime zone created in 1947 took the existing territory of indigenous villages, forest, and farmland and installed a new region that dictated where people lived, what they did, and what opportunities their future held. Such a zone was known as a "post office box" meaning a secret site of military importance. It was also known as "City 40." People experienced radical immobility on entering the zone. The new technical workers were banned from leaving the regime zone without permission of the authorities. For many employees that meant that they stayed put, through family weddings and funerals held elsewhere, through vacations and sick leave. Even as a corpse, workers remained in place, buried in the new settlement's cemetery. Much about the secret settlement was reminiscent of a prison.[36]

The facility was called Mayak (the Russian word for "beacon") and was intended to produce plutonium for nuclear weapons. Though not nearly as well known as the country's nuclear power plant at Chernobyl, Mayak has left a comparable environmental legacy. Releases of radionuclides at Mayak have seriously affected the health of both plant workers and those living in the region and have forced evacuations of entire villages.

Mayak is a vast complex with three main units: a reactor, a radiochemical plant, and a metallurgical plant. Work on the reactor began in September 1946 and it became fully operational in July 1948.

By June 1949, enough plutonium had been obtained to produce the first bomb. This device was detonated at Semipalatinsk now in Kazakhstan on June 29, 1949.

Due to three radiation incidents and normal operation of the plant, significant radiation doses were received in a number of cases by the residents living in the close vicinity of the facility, along with its workers. Cases of chronic radiation sickness were registered within the early period of exposure and in later years increased morbidity and mortality from leucosis and malignant neoplasms were noted.[37]

Problems plagued Mayak's operations from the beginning. In terms of radioactive contamination, the problems at the radiochemical plant were very serious. Due to a design flaw, workers had to clean by hand the filters used to separate out the plutonium and uranium, which exposed them to unacceptably high levels of radiation. Problems with corrosion were also serious leading to the use of materials such as gold, silver and platinum in key equipment. Equipment failures and accidents became common.

Disposing of the plant's radioactive waste also posed a major issue. High-level waste was placed in special storage tanks in an area known as Complex S, while the low level (liquid) wastes were released into the Techa River, which flows through the area around Mayak. As a result of the number of accidents, Complex S was completely full by the end of 1949. At that point, Mayak's management had two options: either stop producing plutonium or dump all the wastes into the Techa River. Because the first option was out of the question in the early days of the cold war, they decided to continue production and attempt to find an acceptable solution to the waste disposal problem. As a result of the activities of the Mayak facility parts of the adjacent territories of the region were exposed to radioactive contamination.[38]

The immediate solution was to use the Techa as a dumping ground. The Americans also dumped radioactive waste in the Columbia River at Hanford. The Columbia, however, was a swift, high-volume mountain river with a rocky bottom, and it raced to the Pacific Ocean. The Techa River was slow moving and turgid. Boris Muzrukov, the boss of the plant at Khystum, wrote a memo that the Techa "is exceedingly polluted."[39] From March 1950 to November 1951, approximately 2.5 mega-curies of radioactive wastes were released into the river. Meanwhile, Mayak's management devised a plan to use nearby Lake Karachay as the principal disposal site. All radiation situations, which occurred in the Urals in the 1950s and 1960s, were caused by inadequate technology for storing radioactive wastes.

The radioactive contamination of the territory in the Urals and the irradiation of the population residing in it were due to the following: releases of radioactive waste into the River Techa; transfer by wind of radionuclides from the dry banks of Lake Karachay and an explosion of a radioactive waste tank in 1957.[40]

The residents of the roadside villages of the river Techa, Iset and Tobol, were exposed both to external and internal radiation. In all, 124,000 people were exposed to radiation on the banks of the river system. The highest radiation doses were received by the residents of the Techa riverside villages (28,000 people) for whom the river represented the main source of their water for drinking and other domestic uses.

Late in 1951, the releases of radioactive waste into the Techa were sharply reduced. In order to minimize the access to the river, protective measures were implemented, including hydrological engineering works: plugging the upper reaches of the river by deep dams, evacuation of the population (over 7,000 people residing mainly in the upper reaches of the

Techa), digging wells, and construction of water pipelines. In order to reduce the effects from the contaminated water on the non-evacuated population, a sanitary protection regime was established which included a fencing-off of the Techa flood-lands at the level of the villages, to prevent the use of river water, using arable flood-lands for agricultural purposes and pasturing, all of which was abandoned. Unfortunately, the resettled population of the riverside villages had already received their main dose of external and internal radiation by the time of the evacuation.[41]

About 124,000 residents in the riverside area of the Techa-Iset-Tobol river system were exposed to elevated levels of radiation. The highest doses were received by 28,000 residents of the Techa riverside, who were exposed to both external and internal irradiation. Between 1953 and 1961, the population of twenty riverside villages (7.5 thousand people) was evacuated.[42]

Soviet scientists engaged in a forty-year study of three generations of villagers who lived along the radioactive Techa River. That secret research followed 1.5 million people including a large control group. By 1986, Soviet researchers had found that people living on the Techa River exposed to chronic low doses of radioactivity had significantly increased death rates and cancers that occurred two to three times more frequently than among Japanese bomb survivors.[43]

Altogether, the operations at Mayak have created five different sources of radioactive contamination: unsafe working conditions and routine accidents at the plant itself; releases of wastes into the Techa River; the storage of wastes in Lake Karachay, the storage of wastes along the Techa and three major incidents including an explosion in a storage tank in 1957.[44] It is on these disasters that we will now focus.

Soviet engineers settled on temporary solutions to dispose of radioactive waste. They built a handful of underground storage containers to store high-level radioactive effluent until a more permanent solution could be found. These engineers dumped low and medium-level waste directly into the Techa River. Soviet scientists knew that the liquids they were dumping exceeded permissible concentrations for an open hydraulic system. Later in 1949, Boris Muzrukov, the boss of the plant at Khystum, learned that the expensive, underground containers of high-level waste had been filled to the brim. Caught unprepared Muzrukov had no place to stash the hundreds of gallons of extremely radioactive effluents discharged from the factory each day. He faced a difficult decision: either stop plutonium production long enough to build more containers or continue operations and find somewhere else to dump the waste products.[45] The immediate solution was to use the River Techa as a dumping ground. From March 1950 to November 1951, approximately 2.5 megacuries of radioactive wastes were released into the river. Meanwhile,

Mayak's management devised a plan to use nearby Lake Kargchay as the principle disposal site.[46]

The second radiation accident occurred on September 29, 1957, at 4:20 p.m. local time in a liquid radioactive storage tank. Although not acknowledged until 1989, there was a chemical explosion at the nuclear fuel reprocessing facility near one of these underground nuclear waste dumps. This accident spread 2,000,000 curies of radiation over 20,000 square kilometers, killed a number of persons, and required the evacuation of thousands. American and British nuclear spokespersons initially dismissed the 1976 report of Zhores Medvedev, an emigré radiobiologist in England, who reported on the accident. However, physicists at Oak Ridge in the United States subsequently analyzed the event and confirmed that it had happened, but they refused to announce this publicly because they feared the growing opposition to nuclear power in their own country. The CIA claimed that the accident was the result of a neutron reactor explosion while producing plutonium. Many persons knew about the accident. The local hospitals were filled with the injured. The authorities posted signs along roads that ran through the highly contaminated regions urging drivers to travel at high speeds and not to stop. Later still, scientists from various research institutes studied the impact of the radionuclides on flora, fauna, soil and water and the pathways through the environment.

The radioactive emissions spread over a region of 3,861-7,722 square miles (10,000-20,000 sq. kilometers). Up to 90 percent of the released activity settled near the damaged storage tank and a radioactive cloud advanced on the provinces of Chelyabinsk, Sverdlovsk, and Tyumen in the downwind direction. Thus, the so-called East Urals Radioactive Trace (EURT) was formed.[47]

The explosion involved concentrated waste produced by military reactors and stored somewhere in an underground concrete bunker. It brought radioactive fission products that had accumulated for years to the surface of the earth to be carried by wind and precipitated dozens of kilometers away by the rain. Although there were a great number of civilian and military laboratories connected to issues of radiobiology, radiochemistry, radio-genetics, and radio-toxicology studying what had happened, issues of secrecy and national security prevented any systematic study of classified materials and precluded publication even when there was such a study. Because of the suddenness of the explosion and the dispersion of the radioactivity, the levels of contamination were only determined after some delay. The secrecy surrounding all this work also hampered timely radiation monitoring.[48] Embarrassment prevented any direct mention of the Kyshtym explosion.[49]

Parts of Chelyabinsk province were exposed to the highest levels of contamination with 10,200 people being evacuated from the contaminated area at different times after the accident.

There were 217 settlements with a population numbering about 270,000 in the contaminated territory. The population of the villages located on the EURT was exposed to the combined radiation from external and internal sources. The external irradiation was due to the increased γ background of the contaminated territories, and the internal exposure resulted from ingestion of radionuclides with locally produced foods. The number of human casualties in such cases is a highly relative question, since radiation damage, especially involving the absorption of radioactive strontium and cesium may not appear in the form of radiation sickness or other pathologies for many months or sometimes years and even decades. The next generation also suffers severely since it is born of parents with an increased burden of strontium 90 in their bones and therefore with a higher level of harmful effects upon the reproductive cells.

In order to reduce the population doses following the 1957 accident a complex of protective measures was put in place. The houses in the villages and settlements were destroyed not by being blown up (in which case the trees would have suffered) but by being burnt down, so that the former inhabitants would not try to retrieve their contaminated belongings.[50] Apart from the evacuation and scheduled resettlements of the population, some foodstuffs and fodder were discarded, a sanctuary control was established, the use of the lakes for agricultural purposes, such as drinking water supply and fishing, was prohibited; the use of pastures and grassland was put under constant radiation control; the state farms had to change their production profile.

The third radiation incident occurred in the spring of 1967, due to resuspension of radioactive material deposited in silt of the drying shoreline of Lake Karachay, an open storage place of liquid radioactive waste. The resulting contamination mainly affected areas of the 1957 trace.

The summer of 1967 was hot and dry, Lake Karachai evaporated and winds blew radioactive dust to areas more than fifty miles away affecting 41,000 people. During the summer of 1967, dust from the lakebed spread over a large area, covering numerous villages, inhabited by over 40,000 people. This was reported to be the most polluted place on the planet and spending a few minutes at this site was predicted to be sufficient to receive lethal radioactive doses. In recent years, efforts to contain the radioactivity were made by filling the lake with blocks of concrete. The consequences on the territory of the 1967 trace were determined by external radiation from the soil due to the settled radionuclides and ingestion via food chains containing local agricultural produce, mainly milk.

The following groups can be distinguished among the populations exposed to elevated levels of radiation in the URALs region.

- Workers of the production association "Mayak" (PA Mayak)

- Population residing in the districts located within the vicinity of the PA Mayak and exposed as a result of the radioactivity emitted during normal operation
- Participants in cleanup and restoration activities
- Residents of the riverside area of the Techa, Iset and Tobel
- Population which lived in the zone of the EURT caused by the 1957 accident
- Population residing in the zone of radioactive contamination resulting from the 1967 downwind transfer of radionuclides from Lake Karachy[51]

Thirty-five years later, on May 20, 1993, the Russian Federation passed a law providing for the social, medical, and political rights of individuals who suffered direct or indirect personal health losses during the liquidation of the accident in 1957 at the Mayak facility or who worked near the Techa River in the period 1949 to 1955 including those who voluntarily or otherwise left the area.

Mayak continued to produce plutonium for nuclear weapons until 1990. There is no doubt that the three Urals disasters were the biggest nuclear tragedy in peacetime that the world has known. It produced the largest radioactively contaminated ecological zone in the world. It will not be gone even a hundred years from now. When people will re-inhabit this region is hard to predict.[52]

For the Soviet Union, the three nuclear disasters at Kyshtum, especially the explosion in 1957, were preludes to what happened later at Chrnobyl, which is the subject of the next chapter.

NOTES

1. Burke, Tony. 2004. Professor Langenkamp, "Nuclear Energy and Proliferation. Problems, Observations and Proposals," Federal Energy and Natural Resource Policy Seminar, December 3, 2004, p. 10. https://lawbepress.com/cgi/viewcontent.cgi?referer=https://www.google.com/&jttpsredir=1&article=4234&context=expresso.
2. Brown, Kate. 2013. *Plutopia: Nuclear Families, Atomic Cities, and the Great Soviet and American Plutonium Disasters.* New York: Oxford, p. 3.
3. Higginbotham, Adam. 2019. *Midnight in Chernobyl: The Untold Story of the World's Greatest Nuclear Disaster.* New York: Simon and Schuster, p. 385.
4. Gephart, R.E., Walker, J. 2005. "A Conversation about Nuclear Waste and Cleanup," *Journal of Environmental Quality,* Vol. 34 (1), Jan. 2005, 399. Hanford, J., Devgan, Jas. 2003. "The Hanford site historic district: History of the plutonium facilities 1943-1990," *Environmental Progress,* Vol. 22 (4), Dec. 2003, pp. D5-D6.
5. Shunya, Yoshimi. 2012. "Radioactive Rain and the American Umbrella," *Journal of Asian Studies,* Vol. 71, No. 2, May 2012, 319-31.
6. Brown, *Plutopia,* p. 6 and 9.
7. Harden, Blaine. 1996. *A River Lost. The Life and Death of the Columbia.* New York: Norton, p. 148.
8. Sanger, S. L. 1989. *Hanford and the Bomb,* Seattle: Living History Press, p. 7.
9. Harden, *A River Lost,* p. 148.

10. Gerber, Michele Stenehjem. 1992. *On the Home Front: The Cold War Legacy of the Hanford Nuclear Site.* Lincoln: University of Nebraska Press, p. 3.
11. US Department of Energy, Hanford Historical Documents 1943-1957, released Feb. 27, 1986.
12. Steele, Karen Dorn. 1988. "Hanford's Bitter Legacy," *Bulletin of the Atomic Scientists,* Vol. 44, No. 1, January-February 1988, p. 18.
13. Gerber, *On the Home Front*, p. 3.
14. Gillette, Robert. 1973. "Radiation Spill at Hanford: The Anatomy of an Accident," *Science,* Vol. 181, No. 4101, Aug. 24, 1973, p. 728.
15. Novikov, Vladimir, Akleev, Alexander and Segerstall, Boris. 1997. "The Long Shadow of Soviet Plutonium Production," *Environment,* Vol. 39, No. 1, Jan. 1, 1997, pp. 12-20.
16. Gillette, op. cit.
17. Medyedev, Zhores. 1980. *Nuclear Disaster in the Urals,* New York: Vintage, p. 151.
18. Delacroix, D., Guerre, J. P., Leblanc, P., and Hickman, P. 2002. *Radionuclide and Radiation Protection Handbook,* Nuclear Technology Publishing ISBN 978-1870965873.
19. Gillette, "Radiation Spill."
20. Medyedev, pp. 148-151.
21. Gillette, "Radiation Spill."
22. Medyedev, p. 151.
23. Ibid, pp. 152-3.
24. Steele, Karen Dorn, "Hanford's Bitter Legacy," pp. 18-19.
25. Higginbotham, p. 385-6.
26. Ibid, pp. 20-21.
27. Radiation Basics, United States Nuclear Regulatory Commission, 2017-10-02, https://www.Nrc.gov/about-nrc.html.
28. Centers for Disease Control and Prevention, Hanford Thyroid Disease Study HTDS Guide, July 31, 2009, https://www.cdc.gov/nceh/radiation/hanford/htdsweb/guide/thyroid.htm.
29. Centers for Disease Control and Prevention, Hanford Thyroid Disease Study (HTDS), Final Report, June 21, 2002, https://www.cdc.gov/nceh/hanford/htds_6-21_public_presentation_for_web.ppt.
30. Steele, Op. cit. p. 22.
31. Harden, *A River Lost,* p. 167; Connor, Tim. 1990. "Nuclear Workers at Risk," *Bulletin of the Atomic Scientists,* Vol. 46, No. 7, Sept. 1990, pp. 24-28.
32. Staff Reports, DOE Dramatically Hikes Estimates for Hanford Cleanup Costs, *Weapons Complex Monitor,* 08 Feb., 2019.
33. Gerber, p. 2.
34. Steele, Op. cit. p. 17.
35. Ibid, p. 9.
36. Brown, *Plutopia,* p. 104.
37. Akleyev, A. V., and Lyubchansky, E. R. 1994. "Environmental and medical effects of nuclear weapons production in the Southern Urals," *The Science of the Total Environment,* Vol. 142, No. 1, pp. 1-8, p. 1.
38. Novikov, Vladimir, Akleev, Alexander, and Segerstahl, Boris. 1997. "The Long Shadow of Soviet Plutonium Production, "*Environment,* Vol. 39, No. 1, January 1, 1997, pp. 12-20.
39. Brown, *Plutopia,* p. 191.
40. Aklyev, A. K., and Lyubchansky, E. R. 1994. "Environmental and Medical Effects of nuclear weapons production in the Southern Urals," *The Science of the Total Environment,* Vol. 142 No. 1, pp. 1-8.
41. Ibid., p. 4.
42. Kossenko, M. M., 1996. "Cancer Mortality among Techa River Residents and Their Offspring," *Health Physics,* 71, 77-82.

43. Brown, Kate. 2019. *Manual for Survival A Chernobyl Guide to the Future*, New York: Norton, p. 35, Collins, Daniel L. 1991. "Nuclear Accidents in the Former Soviet Union: Kyshtym, Chelyabinsk and Chernobyl," Defense National Institute, Uniformed Services University (USU) DNA/AFRRI 4020, ADA 254 669.
44. The Long Shadow.
45. Brown, *Plutopia*, p. 191.
46. The Long Shadow.
47. Lyubchensky, Novikov et alii, p. 5.
48. Medvedev, p. 21.
49. Josephson, Paul R. 2000. *Red Atom Russia's Nuclear Power Program from Stalin to Today*, Pittsburgh, University of Pittsburgh Press, p. 279.
50. Medvedev, p. 22.
51. Novikov, Lyubchensky et alii, p. 6; Present/Science and Technology/Cold War, Science and Space, Nuclear Plant Disasters 2011, *World History Encyclopedia* in ERA9: Promises and Paradoxes, 1945.
52. Medvedev, p. 168.

SIX

Chernobyl

The World's Worst Nuclear Disaster

The explosion in reactor Number 4 at the large Chernobyl Nuclear Power Plant in northern Ukraine, at that time a republic of the Soviet Union, caused a meltdown in April 1986, and this led to an uncontrolled and uncontrollable series of nuclear events. The advocates of nuclear power claim that it is the best option to reduce carbon emissions and supply energy for a growing population. And perhaps nuclear weapons, which are the genesis of nuclear power, are the best way to defend against "rogue" nations, but what transpired at Chernobyl in the wake of the accident at reactor number 4 should give us pause for further thought.

President Eisenhower's concept of "atoms for peace" initially showed promise. The first commercial nuclear power plant was at Calder Hall in 1956 in Great Britain. Nuclear power grew rapidly from 1970 to 1985 but stopped after the Chernobyl accident. Then it began again to increase in number until the Fukushima Daiichi nuclear disaster in 2011 indicated once again the vulnerability of nuclear reactors, this time to earthquakes and tsunamis. Future accidents could have a variety of causes: a failure in personal discipline of the plant operators, an error in the reactor design, or the growing danger of terrorist attacks on nuclear plants. One such attempt was investigated by Belgian authorities in March 2016.[1] Cyber-attacks by hackers constitute another danger, as evidenced by the one that shut down the Chernobyl radiation monitoring systems in June 2017: the Ukrainian authorities believe that this originated in Russia.[2]

On Saturday morning, April 26, 1986, seventeen employees of the Chernobyl Nuclear Power Plant were working on a shift. In carrying out a routine experiment, they turned off the reactor's emergency SCRAM system, which would have been, in any case, too slow to have prevented the accident that followed. As the operators finished the test, they planned to take the reactor off-line for several weeks of routine maintenance. But on shutdown, the chain reaction in the reactor core went "critical" meaning the operators no longer controlled it. The reactor's power surged. Later, in describing the horrific events which followed, the operators remembered how the thick concrete walls wobbled, plaster rained down, and the lights went out. They heard a human-sounding moan as the reactor bolted and then popped. The blast tossed up a concrete lid, the size of a cruise ship, flipping it over to expose the molten-hot core within. A few seconds later, a more powerful second explosion sent a geyser of radioactive gases the equivalent of two hundred Hiroshimas into the splendor of the Ukrainian night.[3] The plant workers had been assured again and again of the safety of the "peaceful atom" and the idea that the reactor had exploded was beyond their imagination.[4]

Firefighters raced to the scene without the assistance of any special equipment or even a clear understanding of the potential risks involved; they had not been trained to deal with a nuclear explosion, because such a training program would have had to acknowledge that an explosion was possible. The efforts undertaken to douse the mysterious, incandescent, fizzy fires only made the conflagrations worse as they seethed with radioactive steam containing uranium dioxide, one of the fuels used in the reactor.

Feeling the heat, the firefighters unbuttoned their jackets and removed their helmets. After less than thirty minutes they began to vomit, developed excruciating headaches, and felt faint and unbearably thirsty. One drank highly radioactive water from the plant's cooling pond, immediately burning his digestive tract. Over the next few hours, the exposed firefighters and plant workers at the scene swelled up, their skin turning the eerie purple of a radiation burn. Later it would turn black and peel away. After evacuation for treatment in Moscow, some of these first responders died and were buried in zinc caskets. Their graves were covered with cement tiles to block the radiation emanating from their corpses.[5]

Helicopter pilots navigated over the disabled reactor, dropping tons of sand, lead and boron to try to snuff out smoldering embers. One helicopter clipped a crane and crashed, killing four men. From the roof of reactor No. 3, soldiers took turns shoveling off the graphite innards of the blown reactor number 4. Miners heroically tunneled ninety feet under the melted core to build a protective wall. Construction workers created dams to hold back the radioactive Pripyat River. Suspecting sabotage, KGB investigators searched filing cabinets, confiscated computer discs, and interrogated survivors dying on hospital beds.

On April 27, more than thirty-six hours after the initial meltdown, army officers escorted 44,500 residents from the atomic city of Pripyat where the workers at Chernobyl had previously made their homes. Columns of Kiev city buses had been sent to wait for evacuees on the outskirts of the city, absorbing radiation while plans were discussed. These now radioactive buses deposited their passengers in villages chosen to house the refugees, then returned to their regular routes despite the fact that they were by then contaminated by radioactivity. In the next two weeks they resettled 75,000 more people from a surrounding thirty-kilometer belt, which was renamed the "Zone of Exclusion"[6] and which remains almost entirely uninhabited to this day. The paradox is that this poisoned zone more than thirty years later is slowly showing signs of great fecundity. The now empty city of Prypiat, whose former inhabitants had provided manpower for the nuclear plant, was becoming invaded by trees and bushes, and the Zone of Exclusion was turning into one of Europe's largest wildernesses. Nature seemed to be thriving in the absence of humanity. Some visitors described it as a kind of radioactive Eden, proof of nature's resiliency.[7]

Radioactive material was dispersed over approximately 39,000 square miles throughout Russia, Ukraine, and Belarus as well as parts of Western Europe and even Canada. On April 28, a radioactive cloud reached Scandinavia, and by April 30 unusually high levels of radioactivity were found in Switzerland and northern Italy. On May 6, the radioactive cloud had reached North America. After attempts at denial, the Soviet government conceded that there had been an accident. To maintain the illusion that this accident was under control, Moscow ordered the Ukrainians to proceed with their planned May Day parade in Kiev, about eighty miles away, thus exposing large numbers of people, including many children, to radioactive fallout.[8] Slowly it became evident that this was the worst nuclear accident in human history, a global problem and a catalyst for the collapse of the Soviet Union.[9]

The accident exposed glaring weaknesses in the Soviet system: its backward technology, its sloppy safety standards, its inability to admit failure. And it brought to the surface many of the injustices, inefficiencies, and secrets that the Soviet government had tried to keep hidden.

By the time of the Chernobyl accident, Soviet citizens had become experts at avoiding accountability for mistakes and failures. Perhaps no other statement quite captures the essence of this lack of responsibility as the one allegedly made by a NIKIET specialist whose organization, the Moscow based Scientific Research Institute of Power Engineering, designed the RBMK reactors at Chernobyl. When asked to aid in the cleanup and mitigation efforts after the accident, he was widely quoted as saying: "This is no longer a nuclear reactor. Our expertise is in nuclear reactors . . . so let others clean it up."[10]

Moscow officials eventually realized that the reactor had exploded and that there was a real chance that another, much larger explosion might occur. A few decades later, the official death toll is now between thirty-one and fifty-four deaths from radiation poisoning (among plant workers and firefighters), doubled leukemia rates among those exposed to exceptionally high radiation levels during the disaster response, and several thousand cases of thyroid cancer. Pripyat became a creepy tourist site.

Reactor Number 4 was not supposed to meltdown. On September 29, 1966, the Soviet Council of Ministers, in Moscow, issued a decree approving the construction of the first of a new generation of giant water and graphite nuclear reactors, which would become known by the acronym RBMK, reactor bolshoy moschnosti kanalny, or high powered channel-type reactor. Developed from the Ministry of Medium Machine Building's military workhorse, the plutonium-and-power-producing Ivan the Second, it was a direct descendent of the pioneering Atom Mirny-I reactor, larger than a two-story house, composed of more than 1,700 tons of moderating graphic blocks and stacked into 2,488 separate columns, each drilled from top to bottom with a circular channel. These channels contained more than 1,600 heat resistant Zirconium-alloy pressure tubes, each of which held a pair of metal assemblies packed with sealed rods of fuel: 190 tons of enriched uranium dioxide, compressed into ceramic pellets roughly the diameter of a man's little finger. Once the reactor went critical and the uranium began to heat, releasing the energy of nuclear fission, the fuel assemblies were cooled by water pumped into the core from below. Under enormous pressure—sixty-nine atmospheres, or a thousand pounds per square inch—the water rose to 280 degrees centigrade and turned into a mixture of water and superheated steam, which was then pumped out through the top of the reactor to giant separator drums. These directed the steam to turbines in order to generate electricity, while the remaining water returned to the beginning of the coolant loop to start its journey through the core once more.

The RBMK was a triumph of Soviet engineering, a testimony to its creators' unrelenting pursuit of economies of scale: twenty times the size of Western reactors by volume, it was capable of producing 3,200 megawatts of thermal energy, or 1,000 megawatts of electricity, enough to keep the lights on for half the population of Kiev.[11] [12]

During the same time frame the Soviets also developed a much safer alternative, the Water-Water Energy Reactor (VVER). The RBMK won out because it generated twice as much energy as the VVER, was cheaper to build and run, and produced plutonium that could potentially be used in weapons. Unfortunately, the RBMK emitted far more radiation than the VVER and had not been fully tested before its operation began. The four reactors at the Chernobyl plant, opened between 1977 and 1983, were all of the RBMK variety. They generated vast quantities of electricity not only for civilian use but also for the voracious energy needs of the nearby secret Duga Radar system located near the plant and which had been built to detect incoming nuclear missiles.

Despite the advanced design of the RBMK reactor, after the Chernobyl explosion an accident containment and mitigation effort began that was portrayed by Moscow as one of the most difficult and heroic engineering tasks ever undertaken. In reality, the period up to the end of November 1986, during which the sarcophagus was constructed, was marred by a poorly performed and reckless effort to conceal the extent of the accident. This was despite the fact that "volunteers," who included former Soviet dissidents and political prisoners, risked their lives in several ineffective accident management actions.

The scale of the damage to Reactor 4 exceeded what the Soviet nuclear authorities believed possible, and they had no contingency plans in place for dealing with it. Most mitigation efforts were devised on the fly.

It was not until midday on Sunday, April 27, that anyone in Moscow had any official idea what had taken place. People had surveyed the remains of the reactor building in the early morning hours of the 26th, but they were either afraid to report what they feared had happened, or they were simply not believed.[13]

After the explosion, burning graphite in the reactor core melted the remaining uranium fuel. The fuel ate through the reactors Lower Biological Shield, flowed into the lower regions of the reactor building, and solidified.

Through most of the first day, Gennady Shasharin, a key Soviet energy official, thought that the core was being effectively cooled by water. Based on this and other incorrect information, the central authorities in Moscow did not immediately sense the urgency of the situation and delayed, for example, the evacuation of residents from what later became the 30-kilometre Exclusion Zone. They said later that they did not want to create panic.

After local firefighters had brought more than thirty small fires under control in the area of Unit 4, efforts were directed at stopping the release of radionuclides from the burning core, which was the principal fire—the now infamous "red glow" thought to be the reactor core—within the reactor building.

Radiation monitors and medical personnel trailed after the Red Army personnel sent to try to assess the damage. Conscripts ripped up asphalt, washed down buildings and removed topsoil with a plan to re-inhabit the abandoned communities. But every time the wind changed, it dusted the landscape with more fallout and the cleanup work would have to begin again.[14]

The Soviet leaders emphasized production over safety. Rather than securing the disaster site and closing it to let the most powerful radioactive isotopes decay for several months or years, they fast-tracked a return of the Chernobyl plant to full production as soon as possible. On May 6, the Soviet authorities officially broadcast to the world that they had extinguished the raging fire in the reactor core. "The danger is over," they announced. This was not true. The fire continued burning until the graphite extinguished itself on its own. Classified records show that radioactive gases poured from the disaster site for another week, spiking on May 11.[15]

Soviet officials estimated that from 3 to 6 percent of the core vaporized into the air and dropped about 50 million curies of fallout on the surrounding environment. A later study conducted after the USSR collapsed estimated that 29 percent of the fuel burned up in the fire for a total close to 200 million curies of radioactivity dispersed into the environment. The released radionuclides were comparable to a very large nuclear warhead, the equivalent of four Hiroshima-sized bombs.[16]

There is, as yet, no accurate guide for societies struggling with large-scale technological and environmental disasters. Reactors, many of which are working long past their expiration dates, are most often built in economically strapped, rural communities where people are grateful for the jobs the plant provides. If a reactor or nuclear bomb factory is shuttered because of an accident or planned obsolescence the immediate territory is abandoned, a cyclone fence goes up, and the radioactive brownfield becomes a nature preserve.[17]

In later years, the most startling revelation made by Chernobyl researchers—and to this day a finding still not fully accepted by the International Atomic Energy Commission (IAEA)—is that the 5,020 metric tons of material dumped from helicopters to smother the smoldering core of the reactor never entered the reactor core shaft. The red glow—ground zero for the helicopter crews—was not the burning core. Rather, it was probably a fragment of the core that had been thrown up and out during the initial explosion. In addition to this bad news, an attempt to quench the fire by displacing oxygen with nitrogen gas most likely occurred after the active phase of the Chernobyl accident had ended.

That these attempts failed suggests that approximately 71 percent of the fuel, or about 135 metric tons, burned virtually in the open for nine days until it melted through the lower part of the core region, flowed downward along piping of the pressure suppression system and quickly cooled and solidified into lava-like substances. All of which means that significantly more radioactivity was released into the atmosphere than the Soviet authorities were willing to admit at the first major Chernobyl conference, the IAEA's Experts Meeting, in Vienna in August 1986.

This does not imply that during the active phase of the accident Soviet officials knew that the helicopter campaign had not covered the core. They almost certainly believed that the helicopter crews had been successful. But by the time of the August meeting, these same officials had had ample time to examine the remains of Unit 4 and to conclude that the core had not been covered.[18]

The scientific finding that the core had not been smothered undermines one of the most important tenets of the official Soviet version of the

Chernobyl cleanup campaign: the cult of the brave helicopter crews who took actions meant to put out the fire, and whose youthful deaths are honored by Soviet authorities by a special museum in Kiev. Clearly, these "liquidators" were brave and selfless, but also misguided. They were also, unfortunately, used by the Soviet authorities to create an impression in the months and years ahead that something had been successfully done to contain the accident.

In August 1986, when academician Valery Legasov, head of the Soviet delegation to Vienna, was faced with the fact that releases of radioactivity began to increase on April 30 and May 1, and that mitigation efforts apparently had been unsuccessful in stemming these releases, he reportedly exclaimed: "The people would not understand. We have to be seen doing something." Later that year, Legasov told the Soviet Academy of Sciences: "I did not lie at Vienna, but I did not tell the whole truth." Legasov committed suicide by hanging himself at home on April 26, 1988, two years to the day after the accident.

In May 1991, Richard Wilson Mallinckrodt, professor of physics at Harvard, spoke at the first International Sakharov Conference on Peace, Progress and Human Rights in Moscow. Based largely on extensive private conversations he had had with Russian scientists, he summarized several ways in which the Soviet government had attempted to control or censor information about the consequences of the accident:

- On Legasov's instructions about six pages concerning radioactivity in Belarus were removed from the official report just prior to the August 1986 IAEA meeting and were not discussed.
- Several pages detailing the large quantities of radionuclides deposited 100 kilometers and more northeast of Chernobyl in the Bryansk oblast of Russia were removed from the report following directives from the Soviet Central Committee.
- Dosimeters in the possession of physicians and private individuals who had worked in the mitigation efforts following the accident were locked up by the KGB.
- Publications of "unauthorized" measurements of radioactivity were forbidden even as late as 1990.
- Physicians in the Ukraine and Belarus were not allowed to mention radiation in their medical diagnoses.
- Appeals by private individuals in Belarus that children not be allowed to drink milk in the first weeks of May 1986 were stopped for fear such warning might induce panic.
- Health records of the "liquidators," soldiers and others who constructed the sarcophagus and did cleanup work in the Zone, disappeared after their work was completed.[19]

The range of Soviet deception regarding Chernobyl seemed endless. The overall effect of the government's propagandistic claims was to draw attention away from the affected people and the extensive contamination of the environment, including great tracts of agricultural land, and focus it on the sarcophagus, which represented "victory" over the accident.[20]

During the summer of 1986, doctors, radiologists and scientists met daily as part of the Chernobyl medical commission. The commissioners struggled with a number of big decisions that would affect millions of people. Soviet authorities had decided to evacuate a thirty-kilometer ring around the plant as a temporary measure. Were there other people who should be moved? Could residents go back to towns and villages inside the circle after the fires had been doused? Repopulating abandoned communities, including the large city of Pripyat, was a priority for the commission. Moscow leaders pushed hard for this because returning people home suggested that the disaster was over, and life could go back to normal.[21]

Dr. Andrei Vorobiev recounted in this memoir that Soviet scientists in the nuclear weapons industry had worked out before the accident that people could safely live on ground contaminated with five curies per square kilometer. They understood this to be a safe number for people who purchased food from elsewhere, but the contaminated ground around Chernobyl was rural. Everyone ate fresh, local produce. Just one curie per square kilometer could deliver unacceptably high doses of radioactivity to a person who consumed local produce. The commissioners looked at the classified radiation maps and decided that if they chose even five curies per square kilometer as a threshold, soldiers would have to eject over a million more people from their homes. Where would they all go? And more importantly, who would pay for it? During those meetings in Moscow, after the fires had been extinguished and crews were building a sarcophagus over the still smoldering reactor, the mammoth scale of the accident suddenly hit home even in far-away Moscow.[22]

Arbitrarily the commissioners selected a new number. They decided that fifteen curies of cesium-137 per square kilometer would be safer or, at least, somewhat safe. Uneasy with this decision the Soviet minister of Health insisted that if people were going to live with fifteen curies, they had to ship in clean food. And what about the 300,000 children and pregnant women, he asked, who were in summer camps for their safety. Should they be allowed to return to areas with fifteen curies?

At their daily meetings in the summer of 1986, the Politburo-mandated medical commissions argued about the doses and how high they could go before they caused unacceptable levels of harm. Today the internationally recognized annual limit is 1 m SV.[23] Representatives from the Ministry of Health drew on the limited data Soviets had of the American-directed Life Span Study of Japanese bomb survivors. With that reference point in mind they suggested that women and children could safely get a hundred times more—100 m SV as an emergency one-year dose. The physicists said a temporary emergency threshold could be much higher—at 500 m SV for the first year, at least for adults.

As the summer progressed, these emergency limits appeared to be more and more optimistic. Soviet researchers determined that people began to have symptoms of chronic radiation syndrome at doses from 200 to 500 m Sv. Biophysicists fought back against the high-dose estimates. Of thousands of people examined, they reported that 90 percent did not exceed 10 m Sv. Only in the most heavily contaminated communities did residents get doses of 300 to 400 m Sv.

The physicians asserted that no one should return to homes inside the thirty-kilometer Zone of Exclusion. They wanted several hundred more contaminated communities evacuated. They sought strict federal guidelines and a definite threshold for further evacuations. The physicists preferred vague formulations and a wait-and-see approach. The physicists largely won the argument.

The new regulations were that in the least contaminated regions (with doses up to 100 m Sv/year), Ilyin wrote, children and pregnant mothers could return home and lead an "unrestricted life." In areas of more contamination, of 100 to 200 mSv per year, residents could return, but they should not eat locally grown food. Forty-five villages, where scientists estimated people would get more than 200 m Sv per year, would need to be evacuated immediately. They determined that places inside the Zone of Exclusion that had levels where people would receive not more than 100 m Sv could be reopened for habitation.

Ukrainian leaders refused to have children and pregnant women return to regions with more that fifteen curies per square kilometer. Children in those areas had already gotten a year's dose, they declared.

Moscow wanted to return residents to eleven Ukrainian villages inside the Zone of Exclusion. Party leaders in Kyiv halted this measure too. The chief sanitation doctor of Kyiv Province, V. V. Malashevsky, had been tracking radioactivity inside the Zone since May. He reported the counts were too high for habitation. Five teams sent from Moscow claimed that the levels were safe. Malashevsky's unit recorded rates ten times higher and declared that to return people was dangerous. The lowly province sanitation doctor did not stand down. Moscow authorities were furious.

The Ukrainian leadership refused to repatriate villagers to homes inside the evacuated Zone of Exclusion. Nor did they fire Malashevsky. In the end, Malashevsky was right. In 1989, those villages were indeed judged once and for all to be too radioactive for habitation.[24]

Across the border in Belarus party leaders faced the same decisions but they did not show the same backbone as the Ukrainians. There are no nuclear power stations in Belarus. Of the functioning stations in the territory of the former USSR, the ones closest to Belarus are of the Soviet designed RBMK type. To the north, the Ignalinsk station, to the east, the Smolensk station, and to the south Chernobyl.

At the end of the summer, the Belarussian leadership obediently complied with Moscow's orders to return children and pregnant mothers to their respective homes and reopen a dozen villages inside the Zone of Exclusion. With that order 1,400 unhappy farmers returned to their villages, seven months after leaving them. Their animals, seeds, preserves, furniture and tools however had disappeared.

They asked for help. Province officials gave them a few sacks of potatoes and told them to stop complaining. Belarussian leaders commanded factory workers and prisoners to go into the Zone to harvest produce and milk. The Belarussian physicist Vasily Nesterenko tried his best to stop the return of several thousand villagers, but alone he could get little traction.[25]

The differences between Ukraine and Belarus could not be starker. For tiny Belarus (population: 10 million) the accident at Chernobyl was a national disaster. As a result of Chernobyl, the country lost 485 villages and settlements. Of these, seventy have been forever buried underground. Today, one out of every five Belarussians lives on contaminated land. This amounts to 2.1 million people, of whom 700,000 are children. Among the demographic factors responsible for the depopulation of Belarus, radiation is number one. In the Gomel and Mogilev regions, which suffered the most from Chernobyl, mortality rates exceed birth rates by 20 percent.[26]

As a result of the accident, 50 million of radionuclides were released into the atmosphere. Seventy percent of these descended on Belarus: fully 23 percent of its territory is contaminated by cesium-137 radionuclides with a density of over 1 Ci/square kilometer. Ukraine on the other hand has 4.8 percent of its territory contaminated, and Russia 0.5 percent. The area of arable land with a density of more than 1 Ci/square kilometer is over 18 million hectares; 2.4 thousand hectares have been taken out of the agricultural economy. Belarus is a land of forests. But 26 percent of all forests and a large part of all marshes near the rivers Pripyat, Dnieper, and Sozh are considered part of the radioactive zone. As a result of the perpetual presence of small doses of radiation, the number of people with cancer, mental retardation, neurological disorders, and genetic mutations increases with each year.[27]

While Ukrainian leaders banned farming in territory with more than 40 curies per square kilometers, Belarussian official permitted people living and farming with counts ranging from forty to a hundred curies of Cesium 137 per square kilometer. Ukrainian villagers received subsidies to purchase food in territories with more than fifteen curies per square kilometer. Until 1988, Belarussians only saw a payout at levels higher than thirty curies per square kilometer.[28]

International agencies and diplomats worked to minimize reports of Chernobyl's damage. The IAEA ended up producing inaccurate estimates of radiation exposure, in part because it had grossly underestimated the use of contaminated local products by people living around the Zone—especially berries, mushrooms, and milk.[29] Despite calls from scientists from many countries, there has never been a large-scale, long-term study of its aftermath. A 1990 assessment by the International Atomic Energy Agency (IAEA) was sabotaged by the FSB (successor agency to the KGB) which was so desperate to conceal sensitive information from foreigners that it stole a large registry of patient data kept on a single computer in Belarus. Ukrainians and Belarussians were told that their health problems were caused by stress and "radio-phobia" rather than by radiation itself.[30]

Red Army pilots were dispatched to seed clouds with silver iodide so that radioactive rain would fall over rural Belarus rather than over urban Russia. Belarussian villagers fell ill, as did the pilots. Livestock absorbed radiation in the immediate aftermath of the disaster by inhaling air and dust, and later by consuming contaminated grass. Cleaned villages were soon re-contaminated by radioactive dust from surrounding areas, and buried material leaked radiation into the water table.

A reluctance to waste food and other basic goods helped keep the radioactive isotopes in circulation. Contaminated wood and peat were burned for fuel in homes and factories, releasing even more radioactivity into the air.[31] Moscow agronomists detailed how to make sausage with an "acceptable" amount of radioactive meat, and Chernobyl sausages were distributed across the USSR without special labelling. There were instructions issued on how to salvage contaminated milk, berries, eggs, beets, grain, spinach, potatoes, mushrooms and tea—often by converting them into products with long shelf lives and simply storing them until the isotopes decayed.[32]

Italy rejected 300,000 tons of radioactive Greek wheat; Greece refused to take it back; the European Economic Community eventually agreed to buy the wheat which was blended with clean grain and sent to Africa and East Germany in aid shipments.

The Chernobyl plant had been built in a marshy area, the worst possible type of land for a nuclear disaster. Mineral-poor soil soaked up radioactive materials, which were then absorbed by mineral hungry plants. Meanwhile, seasonal floods spread contaminants into pastureland.

Radioactivity can move across many environments and exposure to it can come in many variations. We do not even fully understand the process of isotope decay. Biologists originally thought that the ecological half-life of cesium-137 would be only fifteen years; now researchers predict that it will take between 180 and 320 years for cesium-137 to disappear from the forests around Chernobyl though they do not yet know why.[33]

The plume from Unit Four had cast a shadow not only across the cities of Pripyat and Chernobyl but also upon the collective farms and industrial enterprises, small towns, isolated villages, forests, and great tracts of agricultural land. Dense traces of radioactivity reached north and west across the thirty-kilometer zone, but fallout composed of twenty-one different radionuclides forged inside Reactor Number Four—including strontium 89 and strontium 90, neptunium 239, cesium 134, and plutonium 239—had also formed a leopard spot pattern of intense contamination up to three hundred kilometers from the plant. The threat to the population from the radiation was twofold: external, from the fine irradiated dust and debris on the ground around them; and internally, for radioisotope poisoning the food chain through the soil, crops and farm animals. By the end of May more than five thousand square kilometers of land had become dangerously contaminated. Wind and weather would only make matters worse, and dust from radioactive areas was being constantly redistributed into areas the troops had already cleaned, making early decontamination work almost entirely pointless.[34]

Radioactive aerosols had seeped into concrete, asphalt, metal, and wood. Buildings, workshops, gardens, shrubs, trees and lakes had all been in the path of the cloud that had drifted across the landscape for days and weeks. Roofs, walls, farmland, machinery, and forest would have to be washed, scraped, picked clean, or cut down and buried. The reality was that radionuclides could be neither broken down nor destroyed-only relocated, entombed, or interred, ideally in a place where the long process of radioactive decay might pose a less immediate threat to the environment.

Tom Mousseau and Anders Møller are two biologists who have been studying the ecology of the Chernobyl Exclusion Zone since 2000. They have done a great deal of scientific work in the infamous Red Forest, the 1.5 square miles of pine forest which took the largest blast of radioactivity in the days and weeks following the 1986 explosions. It is called the "Red Forest" due to the red-brown color of the dead, irradiated trees.[35] Soon after the accident the pines in the forest, very sensitive to radioactivity, turned red and died; hence, the name Red Forest. Over thirty years, the trees that were planted after the accident acted as cleansing agents, taking up radioisotopes and storing them in their fibers. Normally the Red Forest sent off gamma rays measuring 50-100 μ Sv/hr, but a fire had swept through the forest the previous fall. The hot flames of the forest fire denatured the wood, turning it into gas and ash, and released radioactivity.[36]

Chernobyl clouds missed some areas in the Zone, which today have naturally low levels of background radiation. Other spots are extremely radioactive, places such as the Red Forest, where a person should not linger long.[37] Closest to the power plant, flora and fauna were significantly contaminated. Animals in areas with the most contamination died or stopped reproducing; animals in less affected areas continue to be born with deformities or abnormalities to this day.[38]

Fruit flies have giant chromosomes and reproduce quickly so they make perfect subjects to trace genetic mutations. In the Chernobyl Zone researchers had trouble finding Drosophilia (fruit flies). This is because fruit flies feed on fruit. Fruit trees—apple, pear, rowanberry, wild rose—in highly contaminated areas produced less fruit. Few bees, butterflies or dragonflies survived the accident. The pollinators had been decimated by the release of radioactivity in soils where the insects lay their eggs. Fewer pollinators meant less productive fruit trees. With less fruit, fruit-eating birds like thrushes and warblers suffered demographically and declined in number. Frugivores serve as seed spreaders. With a decline in frugivores, fewer fruit trees and shrubs took root and grew. In nineteen villages in a fifteen-kilometer circle around the blown plant an investigative team found that just two apple trees had seeded in two decades after the 1986 explosion.[39]

In the Red Forest, most of the new growth was in the form of birch trees, which grow better than pine trees because they secrete radioactivity annually when they shed leaves. The pines that did take root were more like shrubs than the tall, straight trees normally grown for board lumber. The floor of the Red Forest had little vegetation. The forests did not smell like forests due to the smell of decomposition. The ground was littered with pine needles and fallen leaves that had not decomposed because the microbes, fungi, and insects that drive the process of decay also suffered from contamination. Even twenty years after the accident, trees in the Red Forest had scarcely decayed. Decomposition, from death to life, is the foundational rule of forest science, a rule violated in the Red Forest.[40]

In 1974 the biophysicist Aleksandr Marei published the results of a four year study—*Global Fallout of Cesium-137 and Man*. Marei's team found that the swampy, sandy soil of the Pripyat Marches were the most conducive of any soil type for transmitting radioactive isotopes into the food chain. Swamps in conditions of continual resaturation accumulate peaty soils that are rich in organic substances but poor in minerals. Plants searching for potassium, iodine, calcium, and sodium readily take up radioactive strontium, and cesium and iodine that mimic these minerals. Marei found that the indigenous berries, mushrooms, and herbs of the marshes showed a very high transfer coefficient of radioactive nuclides from soils to plants. As the boggy soils delivered radionuclides to plants, grazing farm animals magnified radioactive elements in the milk they produced. For Marei the pathway was clear: water, soil, plants, animals, milk, humans.[41]

While world attention focuses, rightly so, on the need to ensure the nonproliferation of nuclear arms, an equally great danger looms from the management of the "atoms for peace." The story of Chernobyl and of Fukushima Daiichi in the chapter that follows point to the need to strengthen international control over the construction and exploitation of nuclear power stations.

Safety procedures at nuclear power stations have indeed improved, the old RBMK reactors have been decommissioned, and new reactors offer a level of safety that nuclear engineers could only have dreamed of in 1986.

Nuclear power seems to provide an easy way out of the growing demographic, economic and ecological crises. But does it? Most new reactors under construction are being built outside the Western world, which is known for the relative safety of its reactors and operating procedures. Twenty-one new reactors are being built in China, plus nine in Russia, six in India, four in the United Arab Emirates, and two in Pakistan. Five new reactors are being built in the United States, and none in Britain. After the Fukushima Daiichi accident Germany under the leadership of Angela Merkel has phased out 8 of its 17 reactors and has pledged to close the rest by the end of 2022.

The next great nuclear power frontier is Africa. Volatile Egypt is currently building two reactors, their first in history. Are we sure that all these reactors are sound, that safety procedures will be followed to the letter, and that the autocratic regimes running most of these countries now seeking to have reactors will not sacrifice the safety of their people and the world as a whole to get extra energy and cash to build up their military, ensure rapid economic development, and try to head-off public discontent? That is exactly what happened in the Soviet Union back in 1986.[42]

The Chernobyl accident led the Soviet authorities to follow a number of procedures they worked out in 1957 at the Ozersk accident thirty-seven years earlier. The use of military conscripts to deal with the consequences of the nuclear explosion, decontamination techniques that included burying contaminated equipment and covering contaminated areas of the nuclear plant with concrete, mass resettlement of the affected population, the creation of an exclusion zone and the handling of large numbers of patients with symptoms of acute radiation poisoning—all of these strategies had first been employed in Ozersk. Original silence about the Chernobyl accident at home and abroad also followed the Ozersk pattern.[43]

The immediate cause of the Chernobyl accident was a turbine test that went wrong. But its deeper roots lay in the interaction between major flaws in the nuclear industry. One such flaw at Chernobyl was the military origin of the nuclear power industry. Chernobyl-type reactors had been developed from the technology that had been created to produce nuclear bombs.[44]

The meltdown at Chernobyl contaminated 40 percent of the European landmass, northern Africa, and North America.[45] Large parts of Europe will remain radioactive for hundreds of years. The hope was that Chernobyl was a unique disaster that would not repeat itself, but then we suffered the meltdown of not one reactor but three at Fukushima Daiichi. This is the subject of the next chapter.

NOTES

1. Shellenberger, Michael. 2018. "If nuclear plants are so vulnerable to terrorist attack, why don't terrorists attack them?" *Forbes*, 6 July. After raiding an apartment owned by a suspected terrorist investigators found images of a nuclear researcher in Belgium. The Belgium Government determined that the threat was "to the person in question, not the nuclear facilities."
2. Plokhy, Serhii. 2018. *Chernobyl The History of a Nuclear Catastrophe*, New York: Basic Books, p. 348. "Petya" cyber attack: Chernobyl's monitoring system hit by world-wide hack", *Independent*, https://www.independent.co.uk/.../ukraine-petya-cyber-attack-hack-nuclear-power-plant-danger-latest-a7810941.html
3. Brown, Kate. 2019. *Manual for Survival. A Chernobyl Guide to the Future.* New York: Norton, pp. 15-16, Evidence of Yuri Tregub, Shift Foreman at Block N4, http://igpr.ru/library/dejstvija_personala_chast_1.
4. Pinkham, Sophie. 2019. "The Chernobyl Syndrome," *New York Review of Books*, April 4, p. 19.
5. Pinkham, Chernobyl Syndrome, p. 22.
6. Brown, Kate. *Manual for Survival*, p. 7.
7. Yablokov, Alexey, V., Nesterenko, Vassily B., Nesterenkol, Alexey. 2009. *Chernobyl Consequences of the Catastrophe for People and the Environment*, Annals of the New York Academy of Sciences, Volume 1181, Boston: Blackwell, 323. The appearance of wildlife flourishing in the Zone of Exclusion at Chernobyl is deceptive. According to morphogenetic, cytogenetic, and immunological tests, all of the populations of plants, fishes, amphibians, and mammals that were studied were in poor condition. This zone is analogous to a "black hole"—some species may only persist there via immigration from uncontaminated areas.
8. Pinkham, Chernobyl Syndrome, p. 22.
9. James, Geoffrey. 2019. Review of *Growth and Decay: Prypiat and the Chernobyl Exclusion Zones* by McMillan, David, *Border Crossings*, Vol. 15, p. 91.
10. Sich, Alexander R. 1996. "Truth was an Early Casualty," *Bulletin of the Atomic Scientists.* Vol. 52, No. 3, 01 May, 1996, p. 32.
11. Higginbotham, Adam. 2019. *Midnight in Chernobyl The Untold Story of the World's Greatest Nuclear Disaster.* New York: Simon and Schuster, pp. 60-61.
12. Pinkham, *Chernobyl Syndrome*, p. 23, Kate Brown, *Manual for Survival*, p. 137.
13. Sich, Truth, p. 32.
14. Kate Brown, *Manual for Survival*, p. 7.
15. Ibid, p. 8.
16. Wilson, Richard. 1987. "A Visit to Chernobyl," *Science* 236 No.4809, 1636-40, at 1637.
17. Kate Brown, *Manual for Survival*, p. 10.
18. Sich, Truth, p. 32.
19. Ibid, pp. 34-35.
20. Ibid, p. 35.
21. Kate Brown, *Manual for Survival*, p. 67.

22. Ibid, p. 68.
23. The rem has been replaced by the standard international unit, the Sievert (Sv) and its smaller subunit, the milisievert (mSv).
24. Kate Brown, *Manual for Survival,* pp. 72-74.
25. Ibid, p. 74
26. Alexievich, Svetlana. 1997. *Voices from Chernobyl.* London: Dalkey Archive Press, p. 1.
27. *Belaruskaya entsiklopedia,* in Alexievich, *Voices,* p. 2.
28. Brown *Manual for Survival,* p. 75.
29. Ibid, pp. 235-6.
30. Pinkham, "The Chernobyl Syndrome," p. 23.
31. Ibid, p. 24.
32. Ibid.
33. Ibid.
34. Higginbotham, *Midnight in Chernobyl,* p. 248.
35. Hutchinson, Tara. 2019. "Chernobyl: Not just another HBO Drama," HeinOnLineBlog, July 15. https://home.heinonline.org/blog/2019/Chernobhl-not-just-another-hbo-drama.
36. Brown, *Manual for Survival,* p. 125.
37. Ibid, p. 126.
38. Hutchinson
39. Ibid, p. 127.
40. Ibid, p. 128.
41. Ibid, p. 134.
42. Plokhy, *Chernobyl,* p. 347.
43. Ibid, p. 174.
44. Ibid, p. 347.
45. Yablokov, Alexey V. et alii, *Chernobyl,* p. 5.

SEVEN

Fukushima Daiichi

The Greatest Industrial Accident in History

The nuclear energy produced at power plants such as the ones at Chernobyl in the Ukraine and at Fukushima Daiichi in Japan offer many advantages to the human community. Electricity generated there is relatively inexpensive considering the amount of energy produced, which means that it can be used in relatively fossil fuel poor nations such as the Ukraine and Japan. This energy does not emit carcinogenic air pollution or carbon dioxide, making it an important component for any nation coming to terms with air pollution and/or climate change. But there are also obvious drawbacks to the technology. The waste generated by the nuclear fuel cycle is extremely dangerous and must carefully be kept away from humans. Nuclear plants, with a remote capability for an accident such as happened at Chernobyl and later at Fukushima Daiichi, and a remote possibility of a terrorist attack, require a great deal of security and bureaucratic oversight, thus making their use very expensive. There is also the relationship between civilian and military applications, posing a number of modern challenges to policy makers. And perhaps the greatest problem is that when things go wrong such as at Chernobyl or at Fukushima Daiichi part of planet Earth is contaminated by radiation and may remain so for thousands of years.[1] If and when an accident does occur we are offered a glimpse at the devastation that would happen in the event of a full-scale nuclear war. According to a news release of the Japanese Ministry of Economy, Trade and Industry on August 26, 2011, the amount of radioactive material released into the environment by the disaster at the Fukushima Daiichi Power Plant was 168.5 times as large as the Hiroshima atomic bomb.[2]

The Fukushima Daiichi nuclear power plant (FDNPP) is located on the Pacific coast of the Tohoku Region in eastern Japan. Until March 11, 2011, the accident at Chernobyl had been the worst nuclear disaster in history, but it involved only one nuclear reactor. There were three reactors which experienced a meltdown at the Fukushima Daichii Nuclear Power Plant. On March 11, 2011, at 2:46 p.m., an earthquake measuring nine on the Richter Scale struck the north east coast of Japan. The earthquake and subsequent tsunami led to the loss of about 16,000 lives, and 4,000 persons were missing. There was also extensive property damage.[3]

Following the established procedure, the Fukushima Daiichi power plant shut down automatically to initiate its cool down. After suffering no apparent damage due to the earthquake, the plant operators breathed a collective sigh of relief. At the moment of the accident, of the six reactors available, only three were operational, units 1 to 3. Unit 4 was completely defueled and units 5 and 6 were shut down for maintenance.

Approximately one hour later however, a tsunami with a 14-meter wave height reached the power plant and caused extensive damage. Specifically, it disabled the emergency power supply thereby interrupting the shut-down procedure, induced a complete black out and destroyed the sea-water pumps of the station.[4] The tsunami that began the nuclear disaster could and should have been anticipated and that ambiguity about the roles of public and private institutions in such a crisis was a factor in the inadequate response at Fukushima. This accident ended the carefully cultivated myth, promoted by nuclear power advocates, of "absolute safety." The crisis was aggravated by dysfunction within and between government agencies and the Tokyo Electric Power Company (TEPCO) in regard to political leadership and crisis management.[5]

Within days the ensuing tsunami crippled the FDNPP.[6] The emerging crisis at the plant was complex, and, to make matters worse, it was worsened by communications gaps between the Japanese government and the nuclear industry.[7] It is clear in the FDNP accident that even in the technologically advanced country of Japan, the government and the plant operator TEPCO were astonishingly unprepared at almost all levels, for the disaster that started with an earthquake and tsunami. And this grave mistake will affect the Japanese people for decades.[8]

This accident was similar to the ones at Three Mile Island Unit 2 on March 28, 1979, and at Chernobyl Unit 4 on April 26, 1986. All three accidents had significant fuel damage and involved the release of fission products into the environment. Fukushima Daiichi and Chernobyl were however, assessed as severe accidents rated as Level 7, or the highest level, events on the International Nuclear and Radiological Scale (INES). The Three Mile Island-2 accident was less severe and was an INES Level 5 event. The total fission product released from Chernobyl was estimated to be about ten times larger than that at FDNPP, and so in that respect Chernobyl was more serious than that at Fukushima Daiichi.[9] At the time of the Chernobyl disaster the nuclear industry hoped that the disaster could be dismissed as being the consequence of a badly designed reactor, not used outside of the Soviet Union and operated in an idiosyncratic way that could not be credible in the West. It was argued that it had little or no relevance to Western designed reactors such as the ones at FDNPP.[10] This was proven to be incorrect.

A few words on nuclear fission might be in order at this point. Nuclear reactions involve the collision of nuclear species and include elastic scattering, inelastic scattering, fission, fusion activation and various particle production events. The fission process has energy production applications because it produces about 200 MeV of thermal energy per event and can be made to be self-sustaining. This is why it is the basis of the thermal power production of nuclear reactors. [11]

Naoto Kan, the Japanese prime minister at the time of the FDNPP disaster, has stated that what took place had two causes. "The first was the total power outage at the Fukushima Daiichi nuclear power plant due to the massive earthquake and tsunami, both of which were unprecedented in the history of Japan. The second was man-made: no one had anticipated such a scenario, and so the government had not taken precautions to build adequate facilities and communications structures."[12]

The earthquake and tsunami that struck the FDNPP disrupted normal and emergency cooling systems. This disruption led to the breaching of three fission product barriers and the subsequent release of radioactive materials from the FDNPP which, thankfully, were primarily fission products with very limited quantities of uranium and plutonium created during reactor operations. When uranium undergoes fission, fission products accumulate within the core of the reactor. Because the fission products are radioactive, they produce heat.[13] These released fission products contaminated air, water, soil, animals, fish, milk and crops, and the resulting dose rates required the evacuation of citizens within 20-30 km of the facility.[14]

Units 1, 2, and 3 at the FDNPP suffered core damage, and their supporting fission product barriers were damaged. Unit 4 was in a maintenance mode and was undergoing re-fueling at the time of the event. All fuel normally residing in the reactor pressure vessel had been moved to the unit's spent fuel pool.

During the earthquake, the external power supply was lost to the reactor complex. The pumps that circulate up to one million gallons of water per minute to cool each reactor core ceased to function. Emergency diesel generators situated below the plants began working, but these were soon shut down by the tsunami. The tsunami breached the protective seawall at FDNPP and disabled the emergency diesel generators. There was no electricity and so the pumps were not operable. Nor could anyone deliver water to cool the reactor cores. After the tsunami made the diesel generators inoperable, only station batteries were available to power the pumps of the emergency cooling system that removed fission product to decay heat. However, this arrangement was only effective until the station batteries were discharged.

After the batteries were depleted, core and spent fuel pool cooling capability were lost, the core temperatures increased in the reactor pressure vessels. Without cooling, the radioactive cores in Units 1, 2, and 3 began to melt within hours. The reactor core consists of sintered uranium ceramic, which does not melt below 2,800 degrees Celsius. The heat in Unit 1, however, was so intense that its core melted. The melted ceramic moved through the steel and into the floor of the containment vessel, the purpose of which was to seal off radiation. The fuel then burned through the protective wall, and radiation was released into the environment. At the same time, the hydrogen generated when the reactor core melted down caused an explosion in the building.[15] Over the next few days, all three cores, each weighing up to one hundred tons, melted through six inches of steel at the bottom of their reactor vessels and spread onto the concrete floor of the containment buildings. Rising pressure necessitated that the steam relief valves be opened to reduce reactor pressure, vessel pressure, and steam which were discharged to the containment vessel. Since there was no capability to add water to cool the core, the water level in the reactor pressure vessel decreased. The core was eventually uncovered, which caused increased heating of the fuel.

When the core temperature exceeded about 1,200 degrees celsius, the remaining water/steam in the reactor pressure vessel produced hydrogen. As the fuel/cladding temperatures increased, steel structures adjacent to the core in the reactor pressure vessel melted, and the fuel and zirconium cladding covering thousands of uranium fuel rods reacted with the water to form an alloy which initiated hydrogen explosions in Units 1, 2, 3, and 4.[16] With severe fuel damage (i.e., fuel cladding damage and melting), fission products were liberated from the fuel and released into the reactor pressure vessel.[17] Massive quantities of radiation escaped into the air and water. The nuclear accident was itself a compound disaster, with meltdowns of reactor cores at Units 1, 2, and 3 and problems with the cooling of spent fuel pools at Units 1 through 4 of the six-unit plant. A hydrogen explosion at Unit 1 on the second day of the crisis exposed a spent-fuel pool to the open air, released radioactive material into the environment and deteriorated the situation at the plant, causing delays in cooling Unit 3. An ensuing hydrogen explosion at Unit 3 then damaged seawater injection lines and vent lines for Unit 2, producing delays in cooling there. In other words, an accident at one unit inevitably hampered responses to the situation at another, leading to parallel chain reactions of accidents and radiation releases. A lot was happening at Fukushima, all at once.

The Fukushima crisis revealed the dangers of building multiple nuclear reactors close to one another. This proximity triggered the parallel chain reaction accidents that led to the hydrogen explosions blowing off the roofs of reactor buildings and water draining from open-air spent fuel pools—a situation that was potentially more dangerous than the loss of reactor cooling system itself. Because of the proximity of the reactors, Masao Yoshida, the director of the FDNPP at the time of the accident, was put in the position of trying to cope simultaneously with core meltdowns at three reactors and exposed fuel pools at three units.[18] In summary, Units 1, 2, and 3 at the FDNPS suffered core damage and their supporting fission product barriers were damaged.[19]

The first twenty-four hours of the accident—from the total loss of power to the station on March 11, when standard cooling stopped, to the forced injection of seawater the following day, were critical. The chairman and the president—TEPCO's two top managers—were not at the head office between Friday, March 11, and 10 a.m. on Saturday, March 12, the most crucial period for dealing with the accident.[20] The most significant mistakes on that first day may have been the misjudgment of the state of the isolation condenser, and the fifteen-hour delay in venting Unit 1's containment area to ease rising pressure.

At the outset of the accident, a TEPCO worker misjudged the backup cooling situation in Unit 1. He failed to notice that the valve of the unit's isolation condenser, or IC, a battery powered emergency cooling system-was either fully or partially closed down after the plant lost power. Steam usually spews out when the IC is activated; because there was no sign of steam, the worker hastily assumed that the IC system had lost its cooling water. For fear of the mechanical damage if the system were run without water, the worker removed the IC from service for about three hours, starting at around 6:30 p.m. on March 12.

Even after top TEPCO official made the decision to vent, there was a more than seven-hour delay, the exact cause of which remains unclear. Whatever the reason for the delay, it resulted in the creation of more hydrogen in the Unit 1 reactor vessel—hydrogen that was vented into the reactor building, blowing off the roof and exposing the reactor's spent fuel pool to the environment.

Many human errors were made at Fukushima. But the role of human error in the Fukushima nuclear accident was not limited to the misjudgment of any one worker, like the one who misjudged the backup cooling situation at Unit 1. The technical chief, the plant director, and the nuclear energy section of TEPCO's headquarters all failed to ascertain the true operational situation of the IC system at Unit 1.[21]

The Nuclear and Industrial Safety Agency (NISA) and the Nuclear Safety Commission (NSC) which oversees NISA's activities as the safety authority—share the blame for the poor response at Fukushima. An egregious oversight, for example, was the fact that the NSC did not include provision for an extended loss of power in its accident-management policy.

At Fukushima, transmission lines for external power sources were not restored until March 17, and the emergency power system could not be quickly repaired. The accident was the result of an extended loss of electrical power, so the regulatory guidelines issued by NSC and executed by NISA which stated that a station black out need not be considered played a large and negative role in the events that transpired.[22]

In hindsight, March 15 turned out to be a crucial turning-point; an early morning accident at Unit 2 led to a dramatic rise in the diffusion of radioactive material from that site. This nullified any hope of containing the radioactivity. The System for Prediction of Environmental Emergency Dose Information (SPEEDI) was developed in 1984 for exactly this kind of situation. The system was intended to help governments decide precisely when to evacuate residents—and from which specific areas. The failure to use SPEEDI suggests that the heavy investment in time and money to develop this system had been wasted.[23]

The Fukushima disaster is not yet over and will not end for many millennia. The radioactive fallout, which has covered vast swaths of Japan, will remain toxic for hundreds of thousands of years. It cannot simply be "cleaned up," and it will continue to contaminate food, humans and animals. It is unlikely that the three Fukushima Daiichi reactors that experienced total meltdowns will ever be disassembled or decommissioned. The IAEA predicts that it will be more than forty years before any progress can be made because of the dangerous levels of radiation at these damaged reactors.[24]

Why were preparations for a nuclear accident so inadequate? One factor was the myth of "absolute safety" of nuclear power. A public relations effort on behalf of the absolute safety of nuclear power was deemed necessary to overcome strong anti-nuclear sentiments connected to the atomic bombings of Hiroshima and Nagasaki. Japan's nuclear community had fears that preparations for a nuclear accident would in themselves be a source of anxiety for people living near the plants.

One of the most disturbing aspects of Fukushima was that it was beyond "design-basis." In other words, the designers did not believe that such a sequence of events was credible. The 9/11 attacks had clear relevance to nuclear reactors, and, in Europe and the United States, designers now have to cope with the possibility that a large airliner might crash into a nuclear power plant. The design of nuclear plants now has to take into account the risk of an earthquake/tsunami.

The Fukushima Daiichi reactor was designed by General Electric and built by Electric Bond and Share Company (Ebasco). Their engineers made six critical design errors in 1965 that were to doom Japan in 2011.

1. The height of the cliff where the plant was located was reduced from thirty-five meters to ten meters.
2. The tsunami wall was too short.

3. The diesel generators were placed in the basement.
4. The emergency pumps located on the shoreline were not submersible pumps.
5. The diesel fuel tanks were placed on the floodplain.
6. The flawed Mark 1 containment was unable to contain radiation.[25]

The first five critical mistakes came from the American engineers' ignorance of the power of a tsunami. Japan has more earthquakes than anywhere else in the world, and its nuclear industry is affected by them in many ways. Tsunamis, which are produced by a seaquake or underwater volcanic eruption, can travel thousands of miles across open ocean at speeds of 950 kilometers per hour and a height of up to 50 meters. The 2011 tsunami was "only" 15 meters high.[26]

After Fukushima Daiichi was struck by the earthquake and tsunami, the nuclear plant lost its ability to generate electricity and to draw electricity from the power grid. The diesel generators intended for emergency use were flooded by the tsunami. But the radioactive materials in the reactor core continued to produce heat. Without cooling, the reactor core would melt down. Cooling required water, delivering water required a pump, and operating a pump required electricity. But there was no electricity, and the pumps were therefore inoperable.[27]

Given the problems in the original design of the FDNPS reactors TEPCO certainly failed to put into place adequate tsunami counter measures, such as reinforcing the embankment and protective system for cooling the water pumps.[28]

Having said all this, during the actual emergency at FDNPS it would appear that key actions were taken by the Japanese authorities during the event should go a long way toward mitigation of any short-term radiologic health impact. These actions included an organized evacuation of over 200,000 inhabitants from the vicinity of the site early in the emergency; monitoring of food and water, distribution of potassium iodide (KI) and systematic screening of evacuees.[29] However, the long-term environmental and health impact will likely take years to more fully understand.

It is important to note that of 295,211 children under the age of eighteen who have been examined by thyroid ultrasound in the Fukushima prefecture, eighty-nine have been diagnosed with thyroid cancer and forty-two more are suspected of having the disease.[30] In Chernobyl, thyroid cancers were not diagnosed until four years after the accident. This early manifestation of thyroid cancer at Fukushima indicates that Japanese children almost certainly received a high dose of radioactive iodine.[31]

The Japanese government is doing its best to clean up the radioactive contamination from Fukushima Daiichi. But in reality, all that can be done is to collect it, place it in containers, and transfer it to another location. The main question becomes where to store the contaminated material safely away from the environment for thousands of years. No container remains effective for longer than one hundred years. Sooner or later they will leak long-lived radioactive elements. There is no safe place in Japan to store this amount of radioactive soil and water.[32]

There was a six-day delay before the radiation emitted from the FSDNPP reactors reached Tokyo. There, radioactive contaminations were reported in milk, vegetables, and other food. For a period, it was announced that tap water in Tokyo was contaminated and even dangerous, especially for infants. It is assumed that radioactive material in the FDNPP plume that spread into the Tokyo metropolitan area on March 16 and March 22 was deposited on the ground via rainfall. [33] Field measurements revealed that the levels of iodine-131 and caesium 137 increased not only in the nearby regions such as Tokyo but also in other countries and even other continents. Two weeks after the accident at the FDNPP I-131, CS-137 and CS-134 were measured in two different locations in Tenerife (Canary Islands). It is suggested that the radioactive plume that arrived at the island was associated with low altitude air masses. This means that radioactive pollution even reached remote regions such as the Canary Islands.[34]

In order to evaluate the consequences in the terrestrial environment in France after the releases from the damaged reactors of Fukushima Daiichi, the Institute for Radiological Protection and Nuclear Safety (IRSN) carried out a monitoring program involving the sampling of grassland vegetation, and milk and meat from outdoor grazing animals, between April and November 2011. This field data allowed the evaluation of the timing of the transfer and the accumulation of Cs over time in milk and meat in the real conditions of the radioactive accumulation. The first Cs detections were observed in vegetation and milk five to seven days after the detection in the air. In meat the first detections were observed by mid-April.[35] The corresponding effective dose to the local public there was a value less than one millionth of the annual limit for the general public. Therefore, there was no risk to the general public. Trace amounts were also reported from both the east and west coasts of the American continent and from Australia.

Radionuclide (I-131, CS-134, and CS-137) concentrations of monitored freshwater fish—species collected from different habitats (rivers, lakes, and culture ponds) in the Fukushima Prefecture during March 2011 to December 2014 were analyzed. Low concentrations of I-131 were detected from only eleven samples collected in March—June 2011 demonstrating that I-131 transferred to freshwater fish after the accident was not intense.[36]

Although no contaminated food was reported outside of Japan since April 2011, there are serious concerns about the future effects about the radioactive debris that have entered the Pacific Ocean.[37] There are two pathways that radioactive fallout from the FDNPP accident can reach the Pacific Ocean: direct release and atmospheric fallout. Cs 137 was detected at most of the stations of recent studies and Cs 134 was also discovered at most of these stations in the North Pacific Ocean. Before the FDNPP accident Cs 137 was already in the ocean. This originated from the nuclear weapons tests conducted in the late 1950s and 1960s. After the FDNPP accident both Cs 134 and Cs 137 containing the distinctive signature of fallout from the FDNPP disaster have been observed in a wide area in the North Pacific.[38]

Timothy Mousseau continued his studies of the biological effects of the accident at Chernobyl and Fukushima. He found that most organisms he and his colleagues studied showed increased rates of genetic damage in direct proportion to the level of exposure to radioactive contaminants. Many organisms showed increased rates of deformities, developmental abnormalities, eye cataracts, and even tumors and cancers. In studying fourteen species of birds found both in Chernobyl and at Fukushima he and his colleagues found a stronger negative response to radioactive contamination in the Fukushima prefecture than at Chernobyl.[39] Reduced fertility rates were also common. This does not bode well for the future of birds near the FDPP. About 40 percent of male birds in the more contaminated areas of Chernobyl are now completely sterile.[40]

In Fukushima, population censuses of birds, butterflies and cicadas suggested that the numbers of these species were negatively impacted by exposure to radioactive contaminants while other groups (e.g., dragonflies, grasshoppers, bees, spiders) showed no significant declines, at least for the first summer following the disaster.[41]

The radionuclide contamination from the disaster at the FDNPP caused harmful effects at physiological and genetic levels of the common lycaenid butterfly, the pale grass blue Zizeeria maha (Z maha), which is common throughout Japan. As the authors of this study of this butterfly point out we cannot evaluate the effects on humans from the results of their Z maha study. But we need to acknowledge that we are just beginning to understand the effects of long-term, low dose radiation on organisms. At this point no one can say whether or not exposure to this level of radiation is safe or unsafe for humans, not to mention humans with physical or mental weaknesses and those who will be born in the next generation, but we need to be cautious and aware.[42]

NOTES

1. Burke and Langenkamp. 2004. "Nuclear Energy and Proliferation, Problems, Observations and Proposals," Federal Energy and Natural Resource Policy Seminar, December 3.
2. Taira, W., Nohara, C., Hujama, A., Otaki, J. M. 2013. "The Fukushima nuclear accident and the pale grass blue butterfly: evaluating biological effects of long-term, low dose exposures," *BMC Evolutionary Biology*, 13: 168, p. 23.
3. Aoyama, Michio, Tsumune, Daisuke, Hamajima, Yasunori. 2013. "Distribution of 137 Cs and 134 Cs in the North Pacific Ocean: Impacts of the TEPCO Fukushima

Daiichi NPP accident," *Journal of Radioanalytical and Nuclear Chemistry,* Vol. 296(1), p. 535.

4. Yanez, J., Kuznetsov, M. Soto-Iglesias, M. 2015. "An analysis of the hydrogen explosion in the Fukushima Daiichi accident," *International Journal of Hydrogen Energy,* Vol 40 (25), 6 July 2015, p. 8261.

5. Funabashi, Yoichi, Kutazawa, Kay. 2012. "Fukushima in review: A complex disaster, a disastrous response," *Bulletin of the Atomic Scientists,* March 1, 2012, Vol 68 (2), p. 9.

6. Caldicott, Helen (ed). 2014. *Crisis Without End: The Medical and Ecological Consequences of the Fukushima Nuclear Catastrophe* [From the Symposium at the New York Academy of Medicine, March 11-12, 2013] New York: The New Press, p. 1.

7. Funabashi, Yoichi, Kutazawa, Kay, p. 9.

8. Ibid, p. 11.

9. Bevelacqua, J.J. 2012."The Fukushima Daiichi Accident and its Radiological Impact on the Environment," *The Physics Teacher*, Vol. 50 (6), September, p. 356.

10. Thomas, Steve. 2012. "What will the Fukushima disaster change?" (Viewpoint), *Fukushima Daiichi nuclear accident in Japan Report, Energy Policy,* Vol. 45, June, p. 12.

11. Ibid, p. 354.

12. Kan, Naoto. 2014 "No Nuclear Power is the Best Nuclear Power," In: Helen Caldicott (ed), *Crisis without End: The Medical end Ecological Consequences of the Fukushima Nuclear Catastrophe,* from the "Symposium at the New York Academy of Medicine, March 11-12, 2012: New York: The New Press, p. 18.

13. Koide, Hiroaki. 2014. "Living in a Contaminated World," In: Helen Caldicott, *Crisis without End,* p. 22.

14. Bevelacqua, Ibid.

15. Koide, p. 22.

16. Caldicott, p. 2.

17. Bevelacqua, p. 355.

18. Kutazawa, Funabashi, 2012. p. 11.

19. Bevelacqua, p. 355.

20. Ibid, p. 12.

21. Ibid.

22. Ibid, p. 13.

23. Hatanaka, Takashi, Yoshida, Sumito, Ojino, Mayo and Ishii, Masami. 2014. "The Communication of Information Such as Evacuation Orders at the Time of a Nuclear Power Station Accident," *Japan Medical Association Journal,* 57 (5-6), Dec. 2014, 93-319, Funabashi, Kutazawa, p. 13.

24. Caldicott, p. 3.

25. Gundersen, Arnold. 2014. What Did They Know and When? In: *Crisis Without End,* p. 139.

26. Ibid, p. 140.

27. Koide, op. cit., p. 22.

28. Funabashi and Kitazawa, p. 19.

29. Dauer, L.T., Zanzonico, P., Tuttle, R.M., Quinn, D.M., Strauss, H.W. 2011. "The Japanese Tsunami and Resulting Nuclear Emergency at the Fukushima Daiichi Power Facility: Technical, Radiologic and Response Perspectives," *Journal of Nuclear Medicine,* 52, Sept. 2011, 1428.

30. Fukushima Health Management, Proceedings of the 15th Prefectural Oversight Committee Meeting for the Fukushima Health Management Survey, 2014, Fukushima.

31. Caldicott, p. 9.

32. Ibid, p. 11.

33. Ishida, Masanobu, Yamazaki, Hideo. 2017. "Radioactive contamination in the Tokyo metropolitan area in the early stage of the Fukushima Daiichi Nuclear Power Plant (FDNPP) accident and its fluctuation over five years," *PLos ONE,* Vol. 12 (11) p.e. 0187687, Nov. 14, Introduction.

34. Pérez, López, Ramos-López, M., Nayra R. Duarte-Rodriguez, X., Bustos, J. J., Alonso-Perez, S., Cuevas, E., Hernández-Armas, J. 2013. "Arrival of radionuclides released by the Fukushima accident to Tenerife (Canary Islands)," *Journal of Environmental Radioactivity,* Vol. 116, Feb., pp. 180-186.

35. Saecy, L., Pourcelot, L., Parache, V., Roussel-Debet, S., Gurriarant, Orjollet, R.D., Leblanc, Renaud, F. P. 2013. "Transfer factors of cesium in milk and meat determined from measurements made in France in 2011 following the Fukushima Daiichi accident," *Radioprotection,* Vol. 48(3), April, pp. 341-354.

36. Wada, Toshihiro, Tomiya, Atsushi, Enomoto, Masahiro, Sato, Toshiyuki, Morishita, Daigo, Izuumi, Shigehiko, Takami Morita, Takami, Kawata, Gojo. 2016. "Radiological impact of the nuclear power plant accident on freshwater fish: An overview of monitoring results," *Journal of Environmental Radioactivity,* Vol. 151, Part 1, Jan., p. 155.

37. Lankarani, K.B. 2011. "From Hiroshima to Fukushima," *Iranian Red Crescent Medical Journal,* 13(8), Aug., p. 529.

38. Aoyama, Tsumue and Hamajima, op. cit., p. 539.

39. Mousseau, Timothy. 2012. "Reply to Andor Pape Møller regarding "abundance of birds in Fukushima as judged from Chernobyl," by Møller et al. 2012., *Environmental Pollution,* Vol. 169, Oct., pp. 141-142.

40. Mousseau, Timothy. 2014. The Biological Consequences of Chernobyl and Fukushima, In: *Crisis without End,* p. 93.

41. Mousseau, Timothy A., and Møller, Anders, P. 2014. "Genetic and Ecological Studies of Animals in Chernobyl and Fukushima," *Journal of Heredity,* Vol. 105, Issue 5, (Sept.-Oct.) p. 704.

42. Taira, W., et al., pp 2-25.

EIGHT

The Greatest De-Proliferation of Nuclear Weapons in the Postwar Period

Avoiding the catastrophe of a nuclear war is not just about preserving people's lives today. It is also about protecting our future, our potential, and the billions of descendants whose lives would be altered forever by a nuclear war. The list of human-caused catastrophic risks is familiar: global warming and other ecological disruptions, pandemics from pathogens engineered by biotechnology, runaway artificial technology.[1] But the most serious danger today remains nuclear war. The United States and Russia have more than 90 percent of the world's nuclear warheads, and when the USSR ended in 1991 there was a serious risk that some of these might have fallen into the wrong hands.

At the time of the disintegration of the former Soviet Union, 27,000 nuclear weapons were scattered about the former Soviet republics. That number was down by 40 percent from the Cold-War Brezhnev-era peak, but they continued to pose a new and serious nuclear danger to world security. First, there was the risk that political authority devolved to the former Soviet republics would be fragmented. This would mean that the nuclear arsenal would be parceled out in ways that would not have been conducive to nuclear stability or to the safe custody of the weapons of mass destruction. Second, there was the danger of seizure, theft, sale, or use of nuclear weapons or their components during the period of transition, particularly if the custodial system were to disintegrate. Third, and most important, there was danger that any weakening of control over weapons, fissionable materials, sensitive components, or know how could result in transfers outside the territory of the new Commonwealth of Independent States, fueling nuclear proliferation worldwide.[2]

In view of the risk to the nuclear weapons in the former Soviet Union, Ashton Carter, formerly Secretary of Defense of the United States, outlined the steps that had to be taken in order of urgency as follows: First, encourage and assist the prompt securing, disabling and removal of the nuclear weapons through Russia. Second, assure the safety and security of Soviet nuclear weapons during the difficult transitional period. Thirdly, address the question of proliferation outside the Commonwealth of Independent States.[3]

About 11,000 of the weapons were thermonuclear strategic weapons, sitting on top of ICBMs, or submarine launched missiles, or loaded on board long-range bombers. All were poised to strike the United States and its allies at a moment's notice. Most of these strategic weapons were attached to their launchers and were equipped with security and release procedures running up to the president of Russia, his minister of defense, and the chief of their general staff. None of the former Soviet republics had ever achieved unilateral release authority. That remained in Moscow.

The other sixteen thousand weapons in the Soviet arsenal posed less of an immediate danger to the United States, but they were, in fact, far more unsettling. This was because these tactical nuclear weapons were portable and hence more easily stolen. They included artillery shells, interceptor missile warheads, nuclear torpedoes, sea-launched cruise missiles, and portable atomic demolition munitions. Many of these lacked even the most rudimentary locks or other safety devices.

During the Cold War, a large number of these tactical weapons had been placed within other states close to the prospective theaters of war. Many lay within the strife torn trans-Caucasus mini republics near Chechnya.

The Strategic Arms Reduction Treaty, initially negotiated by presidents Reagan and Gorbachev at Reykjavik and Geneva in 1986 and 1987, and clarified at various Bush-Gorbachev summit and ministerial meetings in 1989 and 1990, was executed in Moscow on July 31, 1991, just before the attempted coup in Russia. The attempted coup collapsed, discrediting the power of the Communist Party and creating disintegration over the entire former Soviet Union.

On December 1, 1988, the Supreme Soviet adopted new election laws. Three months later republics throughout the Soviet empire were electing legitimate legislative bodies for the first time in seventy years.

After the failed coup attempt of August 1991 most of the Soviet republics started passing declarations of independence. By December 1991, many had elected and installed legitimate presidents. On December 7, 1991, three of these newly elected leaders, representing the three European republics of Russia, Belarus and Ukraine met in a hunting lodge known as the Bison Forest outside of Brest in southwestern Belarus. It took Boris Yeltsin, Leonid Kravchuk and Stanislav Shishkevich only two days to dissolve their ties to the former Soviet Union.[4]

In May 1992, the five nations concerned with the proliferation of strategic nuclear weapons in the outer republics, namely Russia, Ukraine, Kazakhstan, Belarus and the United States, agreed to meet in Lisbon to deal with the problem. Their intention was to create a protocol that would be added to the existing START agreement. As well, the three non-Russian former Soviet republics also agreed to join the Nuclear Non-Proliferation Treaty (NPT) "in the shortest possible time." Those three states also pledged the elimination of all strategic weapons on their territories within the seven-year START horizon.

Earlier President Bush Senior had seized the moment with a radio and television address offering a unilateral withdrawal of all US tactical nuclear weapons from overseas and from the surface ships of the US Navy, in the hope that President Gorbachev would do the same. Bush also announced plans to take US bombers off their alert status, reduce the number of strategic weapons, decommission the multiple independently targetable reentry vehicles, also known as MIRVs, of which much of the US force consisted, and to terminate the development of a new, small, road-mobile ICBM rocket. The Soviet president responded fully during the week that followed. By the end of 1991, the Soviets and Americans had withdrawn almost all of their tactical nuclear weapons from their forward operating bases. The Soviet tactical nuclear weapons within Eastern Europe, the Baltic-states, Armenia, and Azerbaijan and aboard its surface fleet all came home. Some tactical nuclear weapons remained, however, within the Ukraine, Belarus, and Kazakhstan.

So it was that most of the inventory of Soviet nuclear weapons, at least three quarters of them, came to be relocated within Russia itself at the end of 1991. Unfortunately, that still left thousands deployed elsewhere. There were five thousand nuclear weapons within Ukraine alone. With the coming of independence, that inventory made Ukraine the third largest nuclear power in the world. On January 1, 1992, Ukraine had 1,240 strategic warheads sitting on top of long-range missiles; six hundred cruise missiles attached to TU-95 Bear and TU-160 Blackjack bombers based at Uzin and Priluki respectively; another hundred or so gravity bombs were assigned to those same aircraft; and three thousand highly portable tactical nuclear weapons were still scattered about the Ukrainian country side and seaports.

The problem of Kazakhstan was different. Nearly two thousand nuclear weapons remained in Kazakhstan on December 31, 1991: perhaps one thousand, multi-megaton warheads sat on top of the monstrous SS-18 missiles buried in the Kazak desert, another 320 cruise missiles were assigned to the forty TU-95 bombers and in tunnels throughout the land. Beside the warhead legacy, however, Kazakhstan held rooms full of weapons-grade uranium fuel disks, refined for a Soviet submarine fleet that was no longer operational, and it had been home to the principal Soviet test site, Semipalatinsck-21 in eastern Kazakhstan. This was a place that held a vast warren of tunnels, shops, and machinery as well as a legacy of serious pollution and accidents that made all things nuclear most unwelcome among the populace of the Kazakh Republic.

The legacy of the nuclear disaster at Chernobyl made most Belarussians supportive of a reduction of strategic arms and the turnover of nuclear responsibilities to Yeltsin's Russia. President Shuskevich signed the Lisbon protocols. By virtue of its frontier with Poland and the West, however, Belarus also continued to host a large number of tactical weapons.[5] Shuskevich worked to remove the tactical nuclear weapons from Belarus as quickly as possible and ensured that his parliament ratify START I and the NPT in February 1993. He also sought the removal of the eighty-one road mobile SS-25 missiles on a priority basis, agreeing to do so by 1995. These were associated with the road-mobile SS-25 Topol missiles. By November 1996 Russian and Belorussian officials confirmed that all eighty-one of the SS-25 warheads had been returned to Russia. The missiles themselves were removed shortly thereafter.[6]

Since the explosion of the first Soviet atomic bomb in 1949, the Soviet Union reliably maintained control over thousands of nuclear weapons without a single major incident. Ashton Carter and William J. Perry reflected on how this large inventory of nuclear weapons was returned successfully to Russia. "When we took office in 1993, it seemed to us entirely unlikely that Ukraine, Kazakhstan and Belarus would all stay on the path to become nuclear-free states; that Russia would continue to safeguard and dismantle weapons amidst its titanic social upheavals, that somewhere, sometime, there would not be a sale, diversion, or seizure of these weapons by disgruntled military officers or custodians somewhere across the eleven time zones that had been the Soviet empire. . . . But so far, nukes have not broken loose."[7]

The real point in consolidating the nuclear weapons in the former Soviet Union in the successor states was less to prevent proliferation than to forestall terrorism. "It was all called non-proliferation," Linton Brooks, the former US Ambassador and former Under Secretary for Nuclear Safety and Administrator of the National Nuclear Safety Administration (NNSA), clarified, "but the overwhelming majority of it had nothing to do with non-proliferation. It had to do with counterterrorism. If you secure the Russian arsenal, if you secure Russian nuclear material, you've done nothing about proliferation, because if they decide they want to give it to somebody they still can. What you've done is to make nuclear terrorism more difficult. For all that we bragged about very large expenditures on non-proliferation, most of those expenditures were on things that were really security arsenals."[8]

Politicians within the Ukrainian Rada (parliament) were raising objections to the ratification of START I and the Lisbon protocol. Some wanted compensation for the valuable materials within the weapons being exported: the enriched uranium that could fuel Ukraine's nuclear power grid. Others simply wanted to retain Ukraine's nuclear capability. They did not trust the Russians; they did not trust the West. Others pointed out that the SS-24 missiles had been produced within Ukrainian factories. That meant, they argued, that they were Ukrainian; they should stay in Ukraine.

In September 1993, after a half year of debate, the Russians and Ukrainians with US help agreed on a formula known as the Massandra Agreement for Ukrainian participation in the profits of uranium sales from the weapons being returned to Russia. That agreement was repudiated a few weeks later after a confrontational vote in the Ukrainian Rada. On November 18, 1993, the Massandra Agreement was defeated by a 254–9 vote against ratification. Nominally the issue was the ownership of the forty-six SS-24 missiles and their five hundred nuclear warheads. But it was more complex than that.

The first problem was food: the people in the Ukraine were starving. Ukraine was also starving for energy. It had a large industrial base but no oil or gas supplies within its own territory. About one-third of Ukraine's electrical power was coming from nuclear reactors. Ukraine's next problem was the large number of military facilities from the Soviet-era within Ukraine. Getting those Russian troops to "go home" was difficult especially because there was no place for those soldiers to go home to.[9]

Bill Perry became Secretary of Defense of the United States on February 3, 1994. He oversaw intricate negotiations between Ukrainians, Russians and Americans. The challenge was to create a new and definitive agreement on nuclear de-proliferation, one that would stay and be observed until all nuclear weapons were out of the Ukraine. He got the job done in early January 1996. It was William Perry who ultimately would bring about the greatest de-proliferation of nuclear weapons in the postwar period.[10]

One of the most important tools Secretary Perry had in his hand for combatting the potential nuclear proliferation we have been describing was the Nunn-Lugar Cooperative Threat Reduction Program. This program was born in the United States Senate in the fall of 1991. Since then, Congress made available $1.27 billion out of the Department of Defense budget to help the states of the former Soviet Union handle responsibly weapons of mass destruction and cope with the problems related to these weapons.

The program had several broad purposes, which included: (1) consolidation of weapons of mass destruction in safe areas away from areas of conflict; (2) careful inventory and accounting of these weapons; (3) safe handling of them at a time of considerable domestic turmoil in the former Soviet Union; (4) safe disposition of these weapons, as called for by arms control agreements and also by common sense; and (5) assistance in gainfully employing the thousands of former Soviet scientists who knew how to make these weapons of mass destruction around the globe but who often did not know where their next paycheck would be coming from and also were potentially in great demand for employment by rogue nations and terrorist groups.[11]

Yeltsin of Russia and Clinton of the United States met in Moscow to sign the resulting Trilateral Agreement. The United States would provide foreign aid ($60 million via the Nunn-Lugar Cooperative Threat Reduction program) to support the fabrication and shipment of reactor fuel rods to Ukraine and the return of nuclear weapons to Russia. A mini-housing boom was organized within Russia for the benefit of the returning Russian troops. The Russian Federation agreed to send one hundred tons of uranium fuel to Ukraine's nuclear power plants. For its part, Ukraine agreed to return all its strategic nuclear weapons and would commence the disassembly of the strategic missiles and launch facilities located on Ukrainian soil. The parties added mutual security guarantees to protect Ukrainian independence, although those guarantees were not in the form of a treaty as the Ukrainians had hoped.[12] On December 5, 1994, the leaders of Ukraine, Russia, Britain and the United States signed a memorandum to provide Ukraine with security assurances in connection with its accession to the NPT as a non-nuclear state. This is known as the Budapest Memorandum on Security Assurances.

Article 2 of this memorandum states the following: "The Russian Federation, the UK of Great Britain and Northern Ireland and the United States of America reaffirm their obligation to refrain from the threat or use of force against the territorial integrity or political independence of Ukraine, and that none of their weapons will ever be used against Ukraine except in self-defense or otherwise in accordance with the charter of the United Nations."[13]

Kazakhstan was home to 104 ten-warhead "heavy" SS-18 ICBMs, considered the most threatening weapons in the Soviet arsenal, and 40 bombers, carrying just over 1,400 warheads. Although hosting a smaller industrial base than Ukraine, Kazakhstan is the home of a far greater number of nuclear technicians. The Kazakh government might have been firmly committed to denuclearization, but the individuals with the knowledge and training to produce and care for those weapons could be having second thoughts.[14]

The CIA reportedly estimates that some 900,000 people in the former Soviet Union have security clearances to work with nuclear weapons in one way or another including both military personnel responsible for nuclear operations and the employees of the weapons complexes. Of these, 2,000 reportedly have detailed knowledge of the weapons design, and 3,000 to 5,000 more have worked in uranium enrichment or plutonium production.[15]

On the night of November 25, 1994, a US Air Force C-5 aircraft arrived and departed from the Kazakh town of Ust-Kamnogorsk in the greatest of secrecy. This was a project named Sapphire. The aircraft was departing with over a half a ton of weapons grade uranium on board, only a step ahead of Iranian buyers who had been investigating the Ulba Metallurgy Plant two months earlier. Secrecy was essential because President Nazarbayez did not want his own government to know of this deal; he just wanted the material gone. And gone it was, enroute to the Oak Ridge facility for dilution and reprocessing into fuel rods for sale to commercial US power reactor operators.

Twenty years have passed since Ukraine gave up its nuclear weapons in exchange for security guarantees from Russia and the West. Today, Kyiv feels betrayed, and not only by Moscow.

Ever since the Russian annexation of Crimea, Ukraine is more than ever recalling the Budapest Memorandum. This was signed on December 5, 1994 at the summit of the Organization for Security and Cooperation in Europe (OSCE). In it, Ukraine, a nuclear power at the time, voluntarily gave up its nuclear weapons for security guarantees. The United States, Great Britain, and Russia welcomed the decision of the Kyiv regime at the time to agree to the non-proliferation agreement and pledged among other things, to respect the independence and "existing borders" of Ukraine, which included Crimea. The Soviet Union, Belarus and Kazakhstan signed similar memoranda on the same day.[16]

At the time, Ukraine inherited 176 strategic and more than 2,500 tactical nuclear missiles. This seemed to make Ukraine a powerful nuclear armed state. But, according to Leonid Kravchuk, then the president of Ukraine, this was only formally the case. In fact, Kyiv was powerless. "All the control systems were in Russia. The so-called black suitcase with the start button, was in the hands of Russian president Boris Yeltsin."[17]

"Ukraine could have kept the nuclear weapons, but the price would have been enormous," Kravchuk went on to say. Though the carrier rockets were manufactured in the southern city of Dnipropetrovsk, the nuclear warheads were not. It would have been too expensive for Ukraine to have manufactured them on its own.

Additionally, the West threatened Ukraine with isolation since the missiles were supposedly aimed at the United States. Therefore, "the only possible decision" was to give up the weapons, according to Kravchuk.

The Ukrainian missiles were either transported to Russia or destroyed. As compensation, the regime in Kyiv received financial assistance from the United States, cheap energy supplies from Russia and security guarantees that were enshrined in the Budapest Memorandum.

Admittedly, these guarantees were only a formality, since no sanctions mechanisms had been established at the time.

As recently as August 2014, Kyiv was still requesting negotiations with the signatory state based on the agreement, but without success.

Ukraine's foreign ministry spokesman said that Russia's answer was that there was no reason for negotiations. Moscow, as early as a few weeks after the Crimean annexation in March 2014, had rejected all accusations from Kyiv. The "exit of Crimea from the Ukraine" is the result of "complex international processes," Russian officials said. According to Moscow this is unrelated to Moscow's obligations under the Budapest Memorandum.[18]

The USA also denies that it has not fulfilled its obligation. "The Budapest Memorandum was not an agreement on security guarantees," said US ambassador, Geoffrey Pryatt in Kyiv in May 2014. It was an agreement to respect the sovereignty and territorial integrity of Ukraine. Ukraine is convinced that Ukrainian sovereignty and territorial integrity have indeed been violated with the annexation of Crimea by Russia. The Budapest Memorandum, in other words, was respected by neither the Russians nor the West in the case of Crimea. However, Ukraine continues to hold firm to the Budapest Memorandum and President Petro Poroshenko insists that his country was striving for a successor agreement.

The successful relocation of the nuclear weapons from the former Soviet Union to Russia has meant that while the risk of a global nuclear war has gone down, the risk of a localized nuclear attack has gone up.[19] The destructiveness of nuclear weapons is immense. Any use of them would be catastrophic. Today we live with the fear that nuclear terrorism might make use of these weapons of mass destruction, which is the subject of our next chapter.

NOTES

1. Baum, Seth. 2019. Why Catastrophes can change the course of history, Global Catastrophic Risk Institute, http://www.bbc.com/future/story/-20190408-how-catastrophes-can-change-the-path-of-humanity
2. Carter, Ashton B. 1992. "Reducing the Nuclear Dangers from the Former Soviet Union," *Arms Control Today*, Vol. 22 (1), 1 Jan. 1992, pp 10-14, at 10.
3. Ibid.
4. Reed, Thomas C. and Stillman, Danny C. 2009. *The Nuclear Express A Political History of the Bomb and Its Proliferation.* Minneapolis: Zenith, pp. 205-207..
5. Reed and Stillman, p. 204.
6. Ibid, p. 216.
7. Carter, Ashton B. and Perry, William J. 1999. *Preventive Defense. A New Security Strategy for America* Washington: Brookings Institution Press, p. 77.
8. Rhodes, Richard. 2010. *The Twilight of the Bombs. Recent Challenges, New Dangers, and the Prospects for a World without Nuclear Weapons.* New York: Knopf, p. 146.
9. Ibid., pp. 211-212.
10. Reed and Stillman, p. 210.
11. *Dismantling the Cold War US and NIS Perspectives on the Nunn-Lugar Cooperative Threat Reduction Program,* 1997. Ed. Shields, John M. and Potter, William C. Cambridge MA: MIT Press, p. X.

12. Ibid., p. 213.
13. Norris, Robert S. 1992. "The Soviet Archipelago." *Arms Control Today. Arms Control Association* 22(1), January/February, Loose Nukes Special Issue p. 2.
14. Reed and Stillman, p. 219.
15. Sciolino, Elaine. 1992 . "Soviet Brain Drain Poses Atomic Risk US Report warns," *New York Times*, Jan. 1.
16. Pifer, Steven. 2019. "Order from Chaos. Why Care about Ukraine and the Budapest Memorandum," *Brookings*, December 5th.
17. Goncharenko, Roman 2014. Deutsche welle, "Ukraine's forgotten security guarantee: The Budapest Memorandum," 05.12.2014.
18. Ibid.
19. Obama, Barack. 2009. Speech in Prague, Spring, 2009, Richard Rhodes, *Twilight of the Bombs*, p. 297.

NINE

Nuclear Proliferation through Theft, Terrorism, or Neglect

Nuclear weapons in the hands of terrorists are a frightening prospect. A single nuclear device detonated by a state or terrorist group will change our world forever. A surface detonation in a major US city of a five-kiloton nuclear bomb, one-third the size of the Hiroshima blast and a weapon commonly called a "dirty bomb," would destroy most buildings within a several block radius of the target city. Many people within about a mile from ground zero would suffer severe radiation and burn injuries.[1]

The greatest potential sources of nuclear contraband that could fall into the hands of non-state actors are Russia, Pakistan, and countries that possess nuclear research reactors fueled by highly enriched uranium (HEU). With nearly 11,000 nuclear warheads, Russia has the largest arsenal in the world and the most weapons-grade material. Countries occupying what formerly was the Soviet Union have about a third of the world's HEU-fueled research reactors. Pakistan has ninety to one hundred nuclear warheads, is building new plutonium manufacturing capacity, and faces a dangerous and growing insider threat to its nuclear stockpiles.[2]

The most pressing danger is not that terrorists will acquire a working atomic bomb. Except for the case of Pakistan, the security of the nuclear warheads has improved. The insider threat in Pakistan is kept in check somewhat by its nation-state status. It is not difficult to imagine what would happen if a nuclear warhead traceable to Pakistan were detonated in a Western city. The threat of retaliation is still effective.

The real threat of nuclear terrorism stems from the world's growing stockpiles of plutonium and HEU, both of which can be used to make crude atomic bombs.

A recent US-Russian report catalogs nearly 2,000 metric tons of these materials, which are stored in hundreds of buildings in thirty countries under security conditions that range from "excellent to appalling." The combination of fissile material and poor security increases the probability that some of it will end up in terrorist hands.[3]

Security of the nuclear weapons complex, which includes the weapons, the delivery vehicles, the warhead production facilities, and the nuclear storage sites in Russia and beyond remains a serious concern for many. The specter of nuclear terrorism remains especially high after the thus far unsuccessful efforts of Osama Bin Laden and the Al-Qaeda network and the terrorist Japanese apocalyptic sect Am Shinrikyo to obtain one or more nuclear weapons. The worry is that even after the death of Bin Laden Al-Qaeda continues his project to obtain a nuclear device, but also that other groups are trying to obtain material from the nuclear weapons complex.

The following chapter will discuss these concerns but will begin with the important question of possible neglect at nuclear reactors which might be the cause of serious accidents or terrorist theft. We will begin with Africa's first nuclear reactor near Kinshasa in the Congo the security of which is appalling.[4]

The international watchdog dealing with security is the International Atomic Energy Agency (IAEA) with its headquarters in Geneva, Switzerland which has the commission to ensure the safe, secure and peaceful use of nuclear technology. The IAEA has a number of concerns with the apparently lax security in the research reactor in Kinshasa, Congo, an egregious example of the neglect that could lead to enriched uranium falling into the wrong hands.

In the center of the market stalls, among the buyers and sellers, and gridlocked cars on the road out of Congo's capital and into the Kinshasa hills there is nothing to indicate the way to a group of buildings a few hundred yards down a side street.[5] The dilapidated concrete compound is only protected by a low slung rusted barbed-wire fence and an insecure gate locked only by a single padlock. It would be easy enough to slip through a break in the fence but there is no need for that because the main entrance to what is supposed to be one of the best guarded sites in the Congo is often unmanned.

The IAEA has long regarded Kinshasa's experimental nuclear reactor as a disaster in the making either through an accident that might release radiation into the city or because of lax security.[6] There are now three locks to gain access to the main door to the reactor and uranium rods, because years ago the dictator Mobutu Sese Seko handed over a set of keys to a stranger that included the only key available to enter the heart of the atomic plant. That carelessness is blamed for the disappearance of two rods of enriched uranium sometime in the 1970s. The rods, which contained small amounts of enriched uranium, would have been extremely difficult to use in a nuclear device. But during the 1990s, an Italian smuggling ring, which had obtained one of the rods, believed that there might be buyers for the rod that they had in their possession. The purchasers were in fact Italian undercover law enforcement agents who ensnared the smugglers in what became known as Operation Gamma.[7] The one rod involved in Operation Gamma is believed to have turned up in 1998 on its way to the Middle East via the Italian mafia; the other has never been found.

But new locks aside, there is little outward recognition of concern by the world's nuclear watchdog and among western governments at the prospect of Kinshasa's reactor catching the attention of terrorists scouring the globe for the right ingredients for a "dirty bomb."[8]

At the outset, the United States helped create a TRIGA Mark I reactor designed by General Atomic as part of the American Atoms for Peace program and in gratitude to the Congo which had provided the uranium used in the atom bombs dropped on Japan. It cut off the supply of spare parts to the reactor twenty years ago due to the plant's sad decline. Washington has recently tried to persuade the Congo to hand over their ninety-eight bars of enriched uranium stored in triangular rods about 60 cm (2 ft) long and kept submerged in a circular pool underneath a padlocked metal grate in the reactor.

Congo's nuclear plant was installed at the University of Kinshasa in 1958. In 1970 Mobutu stopped the operation of this first reactor in favor of a newer design, the TRIGA II reactor, which reached criticality in 1972. The first reactor was dismantled, and its dozens of uranium fuel rods stored at the site.

The TRIGA Mark II uses ninety-one rod locations in the active core with some unused rods kept in storage. The ninety-one rods are a mixture of fuel rods, control rods or irradiation channels. In 1971, 20 percent enriched fuel rods were shipped from the United States to Zaire in support of the program. The fuel elements for the TRIGA II reactor are cylindrical stainless-steel rods about 28 inches long with a 1.5 inch diameter. The enriched fuel is composed of 20 percent enriched uranium dispersed in a matrix of zirconium hydride. In each rod there are about 280 grams of uranium metal with a zirconium rod down the middle of each uranium fuel rod.[9] There are graphite caps at each end of the rod that serve as end reflectors for the neutrons. The fuel element weighs less than 25 pounds.

The newer reactor was put on standby in 1998 at the insistence of the IAEA, supposedly because of the security risks associated with the war with Rwanda. But the agency had observed the reactor deteriorate for years. Diplomatic sources say that the IAEA feared that an accident could send radiation into the city and contaminate the water supply. Agency officials have been particularly worried that the reactor is built in an area known for its subsidence. Seven years ago one of the walls was pierced by a piece of metal that was variously identified as part of a missile or something that fell from a plane.[10]

There is also concern that the Kinshasa plant could make an easy target for terrorists. While it would be difficult to use the uranium rods to manufacture a nuclear device, they would be useful in building a more rudimentary "dirty bomb."

Notwithstanding these problems the Congo's nuclear scientists hope to restart the reactor so that it can be put to work on a range of uses from medical research to mine prospecting, eight years after it was placed on standby because of war, poor maintenance and especially lax security.

The disappearance of the uranium rods in the 1970s has never been fully accounted for. The reactor's director, Professor Malu Wa Kalenga, has said that the rod recovered from the Italian mafia in 1999 was probably the one stolen from Kinshasa. Many questions remain unanswered however: How did the Italian smugglers obtain the rod? What happened to the second rod? Why did not the IAEA mount a vigorous search for the rods after they first went missing?

A RAND Corporation study of the current state of the Kinshasa reactor gives us reason to be cautiously optimistic.[11] The missing rods themselves are of no use to the person who stole them. He will want to sell them. However, all rods are indexed and labeled. The attempted exchange of illicit rods would raise immediate alarms in the legitimate reactor community. There is no illegitimate reactor community. The only buyer would have to be someone with an alternative use for the rods. Only the uranium fuel has intrinsic value. The uranium could be used in a nuclear weapon. All research reactor uranium is at a sufficient level of enrichment to be used in a weapon.

The uranium metal needs to be extracted from the zirconium hydride matrix. This is a difficult industrial process but can be done. Each rod has only a few hundred grams of uranium metal at 20 percent U-235 enrichment. The spot price for these enrichment units makes the industrial chemistry for extraction uneconomical. To create a nuclear weapon, many orders of magnitude more material would be needed at this enrichment level. That said, in July 1996 Carlo Monteiro, a Portuguese businessman, and others were arrested for allegedly planning to manufacture and sell nuclear bombs from uranium stolen from Zaire.[12] The IAEA's own reports say that the Congo's atomic energy commission has failed to meet every target on security and safety issues, such as radiation protection.

The state of the Kinshasa nuclear reactor and the still missing nuclear fuel rod are, however, the least of the problems that the IAEA and the world community face with respect to contemporary nuclear proliferation through theft, terrorism or neglect.

With the disintegration of the Soviet empire in the late 1990s the economy basically collapsed, thereby creating a security vacuum in Russia and the former Soviet Union satellite nuclear weapons sites and laboratories. Many scientists and those charged with providing security to the former Soviet nuclear arsenal became either unemployed or were compelled to work for diminished wages, oftentimes not being paid at all.[13] A February 2002 report to Congress by the National Intelligence Council (part of the US intelligence community) with respect to nuclear theft from the former Soviet Union notes the following: "We assess that undetected smuggling has occurred, although we do not know the context or the magnitude of such thefts." It goes on to say that many trafficking incidents simply go unreported or sufficient details are not available.[14] The huge nuclear weapons inventory and production complex in the former Soviet Union remains a vast potential source of supply.[15]

A 1998 incident is suggestive of a volatile security environment. Russia's Federal Security Services (FSB—the successor to the KGB) foiled an attempt by staff members of a nuclear facility in Chelyabinsk province to steal some 18.5 kilograms of "radioactive materials" that "might have been used for the production of components for nuclear weapons." Subsequently, a US academic reported being told during a visit to Russia that the material in question was highly enriched uranium. Where the material was headed and who the customers were, if any, is not clear from published accounts. Also unclear is the extent of outflow that might have occurred before the FSB clamped down on the conspiracy. On the other hand, accounts of the incident do not indicate whether the conspiracy went beyond the planning stage, that is to say, whether the participants actually got their hands on the material in question.[16]

Another prominent case was the August 1994 discovery in Munich in an airliner from Moscow of 580 g of MOX (mixed oxide, which can be used as fuel in light water reactors) that contained 300–350 g of near weapons grade plutonium. The plutonium was probably produced in one of three Russian reprocessing plants. In Tangen, near Konstanz at the Swiss border, 8 g of plutonium were found, 99 percent of which were plutonium 239, a more enriched weapons grade plutonium than is normally found in nuclear weapons. Experts believe it was produced for research purposes at Arzamas-16, the nuclear weapons research center in Russia. The theft of naval nuclear reactor fuel under military custody occurred in Murmansk in 1993. The thieves managed to overcome all security precautions. Russian security forces eventually recovered the fuel rods that contained highly enriched uranium and arrested the culprits.[17] There are many other examples of nuclear theft and smuggling that could be offered to the interested reader. One of the most troubling aspects of nuclear theft cases is how weak the controls were that the conspirators had to overcome. In one case in 1993, for example, an individual walked through a large hole in a fence at a naval base, walked to a small shed, snapped the padlock with a metal bar, entered the shed, took several kilograms of enriched uranium, and retraced his steps, without setting off an alarm or encountering a guard. No one noticed what had happened until hours later—and then only because the thief in question had been careless and had left the door of the shed partly open and the broken padlock lying in the snow. The Russian military prosecutor in the case concluded that "potatoes were guarded better."[18]

During the Cold War from 1949–1991, the Soviet Union manufactured 55,000 nuclear warheads.[19] It is within this context that the controversy surrounding suitcase nuclear weapons arose in May 1997 when the former secretary of Russia's Security Council, General Alexander Lebed, now deceased, was debriefing a US congressional panel headed by Pennsylvania Republican Curt Welden. During this session Lebed told the delegation that there were eighty-four one-kiloton suitcase-size nuclear devices that had gone missing from Russia's nuclear arsenal. According to Dr. Bruce Blair of the Center for Defense Information, Lebed made this discovery "after conducting an exhaustive inventory of Russian nuclear weapons in the 1990s."[20] In an interview with *60 Minutes*, an American news program, Lebed not only repeated these claims, but also raised the number to one-hundred. He went on to say that he was unaware of their present location.[21]

The Russian Defense Ministry along with other senior Russian officials denied these claims and went on to assert that no such weapons were ever manufactured. On October 1, 1997, CNN reported that "after repeated denials from top Russian officials, Boris Yeltsen's former environmental safety adviser, Alexei Yablokov, appeared to give some credence to Lebed's claims. Yablokov said the military may not have a record of the portable nuclear bombs which he said were made in the 1970s for "terrorist purposes" for the Soviet KGB, the former Russian secret service and intelligence agency." It is important to note that according to Pennsylvania's Curt Weldon, the Russian government has since conceded that these devices were indeed produced.[22] We know that Soviet suitcase-sized, small nuclear explosive devices or ADMs do or did exist. They were produced most likely for the KGB (now the FSB), Russian Special Forces (Spetsnaz) and possibly for the Soviet military intelligence during the Cold War, and may still exist today.[23] The Middlebury Institute of International Studies at Monterey and the James Martin Center for Non-proliferation Studies examined the available evidence in detail and produced its findings in a report entitled "Suitcase Nukes: A reassessment." The conclusions drawn from this study stated that the likelihood that any portable nuclear devices went missing during the chaos preceding the collapse of the Soviet Union or in the aftermath of its collapse is relatively low. A second conclusion drawn from this study was that in the event that these devices were somehow stolen or confiscated by an unknown entity, their effectiveness would be low to nonexistent due to the presence in the suitcase bomb of the security device called the permissive action link (PAL) and due to the fact that the bombs needed continued maintenance.[24] A permissive action link or PAL is a security device for nuclear weapons. Its purpose is to prevent unauthorized arming or detonation of the nuclear weapon.[25]

Despite this assessment non-state groups exist today with motivation to cause devastating mass destruction and who also have significant financial resources, material and human resources, and finally permissive sanctuary for their operations. Two such groups in the recent past were Aum Shinrikyo and Al-Qaeda. Aum Shinrikyo and Al-Qaeda have demonstrated a commitment unmatched to date by any other organization.

We will begin with Aum Shinrikyo. Better known for its deadly sarin gas attack on the Tokyo subway, Aum Shinrikyo traces its origins to the early 1990s as a religious sect with an apocalyptic ideology. Aum Shinrikyo recruited scientists and engineers to help it acquire nuclear weapons. Aum has an apocalyptic belief structure in which the world is divided into two opposing forces—good and evil. Group leaders such as Shoko Asahara firmly believed they will prevail after the apocalypse and are motivated to trigger the apocalypse because their own salvation depends on fighting and prevailing in the final conflict and eradicating the enemy. The leadership of these movements often relies on a charismatic hold on their members for their authority and encourages the perception of them as messianic.

The Aum leader's beliefs shaped the group's worldview. Shoko Asahara prophesied about nuclear war. He preached that Aum followers would be the only survivors of the coming Armageggon. Asahara believed that Japan was doomed and that a US-Japan nuclear war would trigger this Armageddon. He was obsessed with the prospect of nuclear war. His obsession with nuclear weapons formed the foundation of all his actions. [26]

Asahara's beliefs resonated with his followers for two reasons: there was a strong reaction to the Cold War and the threatening existence of nuclear weapons and the Japanese historical memory of the nuclear devastation during WWII. He was also influenced by the 1991 Persian Gulf War. He claimed that a conspiracy existed among Japan, the United States and the Jews to gain world domination.[27]

Aum recruited highly educated followers, including more than 300 scientists. These scientists had advanced degrees in medicine, biochemistry, biology, and genetic engineering, but none was clearly a nuclear physicist. Many Aum scientists were in fact second-rate academics who had been lured with promises of unlimited funds and the opportunity to conduct original research. One Aum scientist said he joined because he had anxiety fueled by fears of nuclear war.[28] These academic specialists may have driven Aum toward chemical and biological rather than nuclear weapons. According to nuclear weapons builder Carson Mark, a nuclear weapons team would at minimum require a nuclear scientist, a mechanical engineer, an electrical engineer, and an explosives expert.[29]

Many Aum scientists were second-rate. Even though they had graduated from some of Japan's best schools, they typically were not leaders in their fields. Aum gave them a second chance in their careers. Aum did not wait for scientists to simply show up. They actively proselytized a number of graduate students in the hard sciences, promising research funds, state of the art laboratories, and other inducements. Aum's offer was quite attractive because these individuals did not have many opportunities to conduct their own research and writing. In the real world these scientists were not significant, but in Aum, they would be chiefs of the cult's science and technology agency and research what they wanted at an impressive facility with a great deal of money supporting them.

Aum identified Russia as a potential source of nuclear weapons and cult members made numerous overtures to senior officials and scientists. It appears that the group initially hoped that high-level Russian officials might sell them a nuclear weapon in return for a long-term relationship filled with large bribes and perhaps access to attractive foreign technology to which the sect was privy.[30]

Aum's chief mission in Russia was to buy a nuclear weapon. Senior Aum leader Kiyohide's notebooks includes a notation about purchasing a nuclear weapon for $15 million. He devoted considerable time and money to exploring Russia's advanced weapons market. He made eight trips to Russia in 1994. Some Russian cult members worked at their country's leading nuclear research center, the Kurchatov Institute in Moscow.[31] Aum had access to Russian leaders. Asahara led a delegation to Russia in 1992 to meet with senior government officials. These included former Vice President Aleksandr Rutskoy and former Russia Parliamentary speaker Rusian Khasbulatov. The head of Russia's Security Council Oleg Lobov immediately befriended Asahara and gave him a large old building in Moscow. Lobov wanted to start a Russia-Japan university in Moscow but was having trouble raising money for the project until he met Asahara. Lobov reportedly received between $500,000 and $1 million from Aum.

At the outset, after the Saran gas attack in 1995, most Russian officials denied they had helped Aum in any way; however US Senate investigators found photos in Aum publications that showed Rutskoy, Khasbulatov, Basov and Lobov with Asahara. Lobov later admitted that he met with Aum members but said that he was duped by their charitable nature and that the Russian intelligence service did not warn him away from Aum.

Aum also cultivated ties with Russian scientists. Aum targeted Russia's most important nuclear research facility, the Kurchatov Institute, for expertise and technology. The cult was able to recruit at least two Russian nuclear scientists. Asahara held talks with Nikolai Basov, a Nobel Prize winner for physics, in Moscow in 1992. Aum's Moscow branch proposed a meeting with Russia's Nuclear Energy Minister but was turned down.

These connections allowed Aum access to the black market and to various materials and hardware. Aum wanted to use these contacts to obtain sensitive military technologies. But despite high-level connections, plentiful financial resources, and a weak nuclear security regime, Aum was unable to purchase a nuclear weapon. Senior Russian leaders were apparently unwilling to turn over to Aum critical nuclear design information, special fissile material, or a fully assembled nuclear bomb. The relationship with Aum did not outweigh the importance of the protection of Russia's nuclear program. As a result, Aum chose to focus on building rather than buying a nuclear weapon. Group members investigated the mining of uranium ore in Australia and used the Internet to obtain sensitive information in Russia, the People's Republic of China, South Korea and elsewhere.

Fortunately, none of these efforts was successful. Despite the lax security in Russian nuclear facilities and Aum's high-level contacts in the government, Russian officials were unwilling to provide the cult with what it wanted.[32]

Similar efforts to Aum were made by Osama Bin Laden and the Al-Qaeda terrorist network. There is widespread agreement in the US intelligence community that for some time Al-Qaeda has had a strong interest in and has attempted to acquire nuclear weapons. To date, however, no official has stated publicly that Al-Qaeda has actually acquired nuclear weapons.

As the director of the Defense Intelligence Agency told the Senate Select Committee on Intelligence on February 11, 2002, "Al-Qaeda and other terrorist groups are seeking to acquire chemical, biological, radiological and nuclear capabilities." It should be noted, however, that to date there has been no public confirmation by officials that Al-Qaeda has actually acquired nuclear weapons, or indeed any nuclear material necessary to build a weapon.[33]

Bin Laden had abundant motivation for his nuclear quest. In January 1999, Bin Laden said: "Acquiring weapons for the defense of Muslims is a religious duty. . . . It would be a sin for Moslems not to try to possess the weapons that would prevent the infidels from inflicting harm on Moslems."[34] He has also said in March 1998 that "The United States is the biggest terrorist and rogue and it is the duty of every Muslim to struggle for its annihilation."[35]

Bin Laden articulated at least two rationales underlying his attempts to acquire nuclear weapons. First, for Bin Laden, acquiring nuclear weapons was a solemn religious duty because he intended to use them to defend his co-religionists from the "Jews and Crusaders."[36] Bin Laden also sought to inflict the maximum amount of physical damage on the United States. Given their potential power, nuclear weapons were an obvious means to this end. Bin Laden believed that possession of nuclear weapons would deter the United States from using nuclear weapons against him and his followers.[37]

"We have the weapons [nuclear and chemical] as deterrent. I wish to declare that if America used chemical or nuclear weapons against us, then we may retort with chemical and nuclear weapons." He went on to claim that the group already had nuclear weapons.[38]

Al-Qaeda enjoyed formidable financial and other resources. It had in its possession hundreds of millions of US dollars. At the earliest stage, it had sanctuary in Sudan from 1992 to 1996, and then later in Afghanistan. Its organizational capacity included front companies endowed with seemingly legitimate logistical and financial resources. It demonstrated this operational capacity by simultaneously attacking separate targets in different countries.[39]

Al-Qaeda's focus was on acquiring materials to build a nuclear device. To that end he provided funds to Sudan's National Islamic Front (NIF) including the state-owned Military Industrial Corporation (MIC). The cash strapped NIF regime welcomed Bin Laden's financial largess. For his part Bin Laden gained access to a sanctuary that allowed him and his followers to pursue their interest in nuclear weapons. It is likely that the MIC was the vehicle through which Bin Laden attempted to acquire these weapons.[40] Bin Laden made investments in the local economy of Sudan in agriculture, banking and construction. He established dual-use companies for terrorist cover purposes.

Bin Laden employed a top aide and an early member of Al-Qaeda, Mamdouh Mahmud Salim, as his "point-man" on nuclear matters. In late 1993 Salim agreed to attempt to obtain enriched uranium. Al-Qaeda's early efforts to develop nuclear weapons were disappointing. Salim was apparently the victim of a scam involving low-grade reactor fuel.[41] According to one researcher, "intelligence sources now believe that criminals sold Al-Qaeda irradiated canisters purporting to contain uranium stolen from Russian army bases, whereas in fact the contents would have had no military value whatsoever had they been passed on to rogue nuclear scientists."[42]

Al-Qaeda was also subject to numerous other scams involving radioactive material. Osama's weapons buyers were not physicists. According to press accounts, Osama Bin Laden and his associates were tricked into paying for material called "Red Mercury," which they believed to be weapons-grade nuclear material. In the mid-1990s a number of smugglers claimed to have in their possession nuclear material that they referred to as "Red Mercury" but in most instances the material was fictitious or radioactive waste and not weapons grade nuclear material.[43] Salim was finally arrested in Munich in September 1998 and extradited to the United States.[44]

Al-Qaeda's symbiotic relationship with the Taliban enabled it to use Afghanistan as a secure base for pursuing unconventional weapons capabilities.[45] This sanctuary allowed Al-Qaeda to hide its activities from the outside world in relative security. According to David Albright, the regime may have transferred nuclear material and technology to Al-Qaeda.[46] Other sources allege that the Taliban attempted to recruit a "former Soviet nuclear weapons expert from a Central Asian state," but the plot was disrupted by Russian authorities.[47]

Searches of Al-Qaeda facilities after Operation Enduring Freedom have produced little evidence of much progress in the Al-Qaeda nuclear weapons program. Allied military forces uncovered a number of documents. Search of sites suspected of involvement in nuclear weapons development revealed instruction manuals and other documents related to nuclear weapons. Among them was a weapons design for a "superbomb" discovered in Kabul that experts who examined it characterized as unworkable. If someone obtained separated plutonium and built a bomb following this design, it would not function as an atomic bomb. In Albright's judgement, there was no evidence that the group had gone beyond theory; "To create a nuclear weapon," Albright said, "a designer must learn a whole set of manufacturing steps not mentioned in Al-Qaeda's manual and develop confidence in the weapon's design."[48] These documents reinforce the assessment that Al-Qaeda was determined to acquire nuclear weapons, but also sought radiation dispersion devices (RDDs), an easier step up the learning curve. Some documents were likely destroyed or removed by Al-Qaeda, but there was no evidence of weapons or nuclear material.[49] There is some speculation by observers that al-Qaeda was in fact focusing on a simpler gun-assembled design. Overall, the seized documents suggest a rather limited capability. Geoff Hoon, the UK Defense Secretary, announced in November 2001 that "[we} are not convinced at this stage that he [Bin Laden] is capable of producing a nuclear bomb."[50]

Frustrated by scams when attempting to acquire materials, and (perhaps) lacking technical expertise, Al-Qaeda may have then attempted to buy, rather than build, nuclear weapons. At the same time individuals associated with Bin Laden continued to seek out low-level dispersal devices or "dirty bombs."[51]

Since 1999, there have been numerous and unsubstantiated claims that Russian "suitcase nukes" have ended up in Al-Qaeda hands. There is no evidence that any such exchange has ever taken place. Russian President Vladimir Putin and other senior officials vehemently deny that Al-Qaeda obtained any such weapons from the former Soviet Union.[52] Indeed, the Russian government goes further to deny that Al-Qaeda obtained nuclear weapons from any other sources in the former Soviet Union. In a November 10, 2001, interview, President Putin said he was "absolutely confident" that terrorists in Afghanistan do not have Soviet or Russian weapons of mass destruction.[53]

Pakistani press claimed in November 2001 that two Russian atomic demolition munitions (ADMs) or "briefcase" nuclear weapons, each containing "at least 2 kgs of fissionable plutonium and uranium" have reached US shores.[54] A key question is not answered in these press reports. Are these weapons fitted with permissive action links (PALs) that would have to be used in order to detonate the weapons? According to one study, Russian "portable nuclear weapons are equipped with some protection devices (e.g. PALs) making their unauthorized use difficult though not completely impossible." Other reviews of the Russian suitcase nuclear bomb controversy have been skeptical of the claims that these weapons have been lost or stolen and consequently acquired by Al-Qaeda.[55]

In summary, Al-Qaeda was seemingly poised for success in obtaining a nuclear device. It had the motivation, the resources and sanctuary and infrastructure first in the Sudan and then in Afghanistan. Then there was the question of the reportedly lax security in the Former Soviet Union and the existence of a nuclear black market.

And so, the question is: what intervened to preclude their success? Why did Al-Qaeda fail to build nuclear weapons? First of all they lacked appropriate expertise. They may also have lacked access to the needed material which was perhaps more secure than is commonly understood. Building a nuclear weapon represents a formidable challenge for a sovereignless, subnational entity. The management of such a program is formidable, which is one reason why only a handful of states have succeeded in doing this.

This suggests a second question which is why Al-Qaeda failed in its efforts to purchase nuclear weapons. As Aum Shinrikyo also discovered Russian warhead security though imperfect still is more difficult to penetrate that was often thought.[56] Furthermore, acquiring nuclear weapons seems prohibitively expensive since the market for nuclear weapons does not exist.[57] The conclusion is that we know that it is impossible to determine with any degree of accuracy the present nuclear capability of the Al-Qaeda organization.[58]

The Rand Corporation study would seem to be quite correct that stricter international controls over nuclear weapons, materials and expertise would reduce opportunities for terrorists to acquire nuclear materials.

NOTES

1. Albright, David, Buehler, Kathryn, Higgins, Holly. 2002. "Bin Laden and the Bomb," *Bulletin of the Atomic Scientists*, Jan/Feb, Vol. 58, Issue 1, p.22.
2. Lawlor, Maj. General Bruce. 2011. "The Black Sea: Center of the nuclear black market," *Bulletin of the Atomic Scientists*, Vol. 67-6, Dec. 15, p. 673.

3. Global Fissile Material Report. 2013, "Seventh Annual report of the International Panel on Fissile Materials," fissilematerials.org/library/gfmr13.pdf
4. Miklian, Jason. 2013. "Hot Rod. Poking around for uranium inside the world's least secure nuclear reactor," *Financial Post,* June 28th
5. McGrail, Chris. 2006. "Missing keys, holes in the fence, and a single padlock: welcome to Congo's nuclear plant," *The Guardian,* November 23, https://www.theguardian.com/world/2006/nov/23/congo.chrismcgreal
6. Ibid.
7. Daly, Sara A., Parachini, John, Rosenau, William. 2005. Project Air Force (US): Rand Corporation, *Aum Shinrikyou, Al-Qaeda and the Kinshasa Reactor: implications of three case studies for combating nuclear terrorism*, p. 56.
8. McGrail, op. cit.
9. Daly et alii, p. 59.
10. McGrail, op. cit.
11. Daly et alii. 2005.
12. Daly et alii, p. 61
13. Shanty, Frank. 2007. "Has Al-Qaeda Acquired Nuclear Weapons? The State of Existing Knowledge" In: Shanty, Frank and Mishra, Patit Paban. *Organized Crime: From Trafficking to Terrorism* ABC-CLIO, LLC, p. 356.
14. National Intelligence Council Annual Report on the Safety and Security of Russian Nuclear Facilities and Military Forces, Feb. 2002, fas.org.irp/nic/icarussiassecurity.htm
15. Ibid.
16. Tkachenko, Yevgeny. 1998. "FSB Prevent Theft of Nuclear Materials" ITAB—TASS, Dec. 8; "Minatom Says 1998 Theft Involved HEU."2000. Center for Nonproliferation Studies, NIS Trafficking Data Base, Oct. 31.
17. Bluth, Christoph. 2000. "Review of *Smuggling Armageddon: The Nuclear Black Market in the Former Soviet Union and Europe"* by Lee, Renssalaer W. 2000. *Slavic Studies* Vol. 59, No. 3, pp. 690-691.
18. Bukharin, Oleg and Potter, William. 1995. "Potatoes Were Guarded Better," *Bulletin of the Atomic Scientists"* LI, (May/June) p. 48.
19. Norris, Robert S. and Kristensen, Hans M. 2010. "Global nuclear weapons inventories, 1945-2010, *Bulletin of the Atomic Scientists,* 66:4, 77-83, at p.79.
20. Ibid.
21. Shanty, p. 353
22. Ibid, p. 354.
23. Ibid, p. 356.
24. "Suitcase Nukes: A Reassessment," 2002. Middlebury Institute of International Studies at Monterey, James Martin Center for Non-Proliferation Studies, Sept. 23, nonproliferation.org/suitcase-nukes-a-reassessment
25. Bellovin, Steven M. 2005. "Permissive Action Links, Nuclear Weapons, and the History of Public Key Cryptography," Oct. 21, Department of Computer Science, Columbia University, smb@cs.columbia.edu http://www.cs.columbia.edu visited February 2020.
26. Daly, Parachini, et alii, p. 6.
27. Ibid, p. 8.
28. Lifton, Robert Jay. 1999. *Destroying the World to Save It. Aum Shinrikyo, Apocalyptic Violence and the New Global Terrorism,* New York: Henry Holt, p. 128.
29. Carson, Mark et alii, 2006. "Can Terrorists Build Nuclear Weapons?" *Nuclear Control Institute,* Washington DC, accessed at www:// nci.org/k-m/makeab.htm, visited Feb. 2020.
30. Daly, Parachini, p. 12.
31. Lifton, p. 200.
32. Ibid, p. 20.
33. United States of America v. Usama bin Laden. 2003. Indictment 98, Co, United States District Court, south District of New York, p. 7, Vice Admiral Jacoby, Lowell E.,

Defense Intelligence Agency, Statement of the record, Senate Select Committee on Intelligence, Feb. 11, 2003, p. 3, and Tenet, George J. 2002. Director of Central Intelligence Agency, testimony before the Senate Armed Services Committee, March 19. www:odci.gov/cia/public_affairs/speeches/archives/2002/senate_select_hearing_03192002.html.

34. Osama bin Laden. Interview with Time Magazine, Dec. 23, 1998, pbs.org/wgbn/pages/frontline/shows/binladen/who/edicts.html

35. Daly, Parachini, p, 25.

36. Albright, David. 2010. *Peddling Peril, How the Secret Nuclear Trade Arms America's Enemies*, New York: Free Press (Simon & Schuster), chapter 8.

37. Daly, Parachini, p. 27.

38. Mir, Hamid. 2001. "Osama Claims He Has Nukes: If US Uses N-arms It Will Get Same Response," *Dawn*, Nov. 10, Weiner, Tim. 2001. "A Nation Challenged: Al-Qaeda; Bin Laden Asserts He Has Nuclear Arms, *The New York Times*, Nov. 10, section B, p. 4.

39. Daly, Parachini, p. 28.

40. Anonymous, 2002. *Through Our Enemies Eyes: Radical Islam and the Future of America*. Washington DC: Brassesy's Inc., p. 186.

41. Leader, Stefan. 1999. "Osama bin Laden and the Terrorist Search for WMD," *Jane's Intelligence Review*, Vol. 11, No. 6, June, pp. 34-37; Potter, William C. 2008. "Nuclear Terrorism and the Global Politics of Civilian HEU Elimination," *The Nonproliferation Review*, Vol 15 No. 2, abstract.

42. Gunaratna, Rohan. 2002. *Inside Al-Qaeda: Global Network of Terror*, New York: Columbia University Press, p. 11.

43. Daly, Parachin, p. 34.

44. Davies, Hugh. 1998. "Bin Laden Aide Tried to Buy Atomic Arms," *Daily Telegraph* [London], September.

45. Daly, Parachin, op. cit.

46. Albright, p. 23.

47. Hutchinson, Robert. 2003. "The Struggle for Control of Radioactive Sources," Jane's *Intelligence Review*, March, p. 33.

48. Albright, David. 2002. "Al-Qaeda's Nuclear Program: Through the Window of Seized Documents," Nautilis Institute Special Forum, Vol. 47, November 6. Oldsite.nautilis.org/archives/for a/Special-Policy-Forum/47_Albright/html.

49. Daly, Parachin, p. 37.

50. Independent Online, 2001. "British troops now in Afghanistan, says Hoon" IOL, World, 11 November, iol.co.za/news/world/british-troops-now-in-afghanistan-says-hoon-76697.

51. Daly, Parachin, p. 40.

52. Boureston, Jack, 2002. "Assessing Al-Qaeda's WMD Capabilities". *Strategic Insights* Vol 1, issue 7, September 2, Center for Contemporary Conflict, Naval Postgraduate School, http://hdl.handle.net/10945/25379.

53. National Intelligence Council, *Annual Report to Congress on the Safety and Security of Russian Nuclear Facilities and Military Forces, Feb.*, op. cit.

54. Daly, Parachin, p. 41.

55. Smigielski, David. 2003. "A Review of the Suitcase Nuclear Bomb Controversy," *Partnership for Global Security* September, http//www.partnershipforglobalsecurity-archive.org/documents/suitcasenukes090103.pdf; Seed, David. 2019. "Suitcase Nukes", *US Narratives of Nuclear Terrorism Since 9/11 Worst Case Scenarios.* Liverpool, UK: Palgrave.

56. Daly, Parachin, p. 49.

57. Shanty and Mishra, Op. Cit.

58. Ibid, p. 355.

TEN

Nuclear Weapons Testing

Inflicting the Conditions of Nuclear War on Fragile Environments

Today, for the first time since the beginning of the nuclear age, none of the world's nuclear-armed states is conducting nuclear test explosions. After more than 2,000 detonations, the world's nuclear test sites are dormant. This is good news indeed for it has meant that the world community has been steered away from nuclear testing, and the arms race and environmental contamination it produces.[1] The present chapter will provide details about why this decision is so important for the international community and the global environment.

Nuclear weapons testing was conducted in the atmosphere between 1946 and 1980. The testing resulted in exposure of local populations to radiation as a result of fallout or radioactive debris. The nuclear tests were conducted by five nations: United States, the Soviet Union, the United Kingdom, France, and China. They happened primarily at sixteen sites. The sixteen testing sites are located in nine different countries on five continents and Oceana. These tests have contributed nearly all of the radioactive materials released to the environment by atmospheric testing. The sixteen sites are: the Nevada Test Site, USA, Bikini and Enewetak, Marshall Islands (Oceana), Johnston Island, USA (Oceana), Christmas and Malden Island, Kiribati (Oceana), Reggane in Algeria (Africa), Novaya Zemlya and Kapustin Yar in Russia (Europe) Semipalatinsk, Kazakhstan (Asia), and Lop Nor (China). There were large differences in the numbers of tests conducted at each location and in the total explosive yields. In general, the tests conducted earliest led to the highest individual and population exposure.[2]

The declassified history of nuclear weapons testing has shown us how little responsibility leaders took for damage caused by the detonation of the equivalent of 29,000 Hiroshima-sized bombs.[3]

Between 1945 and 1980, bomb designers detonated their weapons with reckless abandon and seemingly without care for their effects on the people or environment involved in their tests. Alaskans, Scotch, Scandinavians, and many others were hit with Soviet bomb tests at Novaya Zemlya. Australians and Pacific Islanders were showered with fallout from French, British and American bombs that blasted Pacific islands. Chinese and Soviet bomb designers joined forces to pollute the interior of Eurasia. The tremendous quantity of radioactive fallout swirled in the atmosphere and came back down to planet Earth in rain and snow. The more precipitation in a place, the more the radioactivity cascaded down.[4] Radioactive fallout from atmospheric testing was dispersed worldwide, resulting in measurable radioactivity on every continent. The contamination of air and foodstuffs resulted in some internal exposure to ionizing radiation among almost all people born since 1945.[5]

The damage of half a century of nuclear bomb testing appears to have left a perceptible impact on human health. The main health concern from exposure to ionizing radiation emitted from radioactive fallout is the development of thyroid cancer among people exposed in childhood. The thyroid gland cannot distinguish between stable iodine, a necessary dietary mineral, from radioactive iodine, such as Iodine 131, produced in a nuclear explosion. The thyroid, most sensitive to radiation at a young age, can receive an internal dose after consumption of contaminated foods, for example fresh milk and other dairy products that accumulate radioactive iodine when dairy animals graze on contaminated pastures.[6] Rates of thyroid cancer in the United States tripled between 1974 and 2013, and this increase cannot be attributed simply to better detection. As global fallout drifted down, mostly in the Northern Hemisphere, thyroid cancer rates grew exponentially. In Europe and North America, childhood leukemia, which used to be a medical rarity, increased in incidence year by year after 1950. Australia, hit by fallout from British and French tests, has the highest incidence of childhood cancer in the world. Almost all of the radiation-related cases of thyroid cancer would be among individuals younger than twenty years during the period 1951–1957.[7] These statistics show a correlation and not a causal link. They do, however, raise many questions. Thyroid cancer among children is the medical canary in the mine.[8]

People have been exposed over their lifetimes to large doses of man-made radiation from medical procedures, nuclear power reactors and their accidents and atomic bombs and their fallout. Few people on earth have escaped those exposures, which are the legacy of nuclear testing.[9]

Since the dawn of the atomic age in July 1945, in order to increase the lethality of their bombs nuclear weapons have been tested on more than

2,000 occasions—in the atmosphere, underground and underwater. The toll on human health has been devastating. Much of the earth's surface has been contaminated at some point with radioactive particles.[10]

Nuclear tests have been carried out at more than sixty locations around the globe, often on lands of indigenous and minority peoples. The tests have irradiated people working on the programs, the downwind and down-stream communities, and in fact the whole global population.[11]

The Soviet RDS hydrogen bomb, known by Western nations as the Tsar Bomba, was the most powerful nuclear weapon ever created. Tested on October 30, 1961, as an experimental verification of calculation principles and multi-stage thermonuclear weapon designs, it also remains the most powerful explosive device ever detonated. The bomb was detonated at the Sukhoy Nos cape of Severny Island, Novaya Zemlya, 15 km from Milyushikha Bay. Soviet scientists in 1991 revealed that the instruments indicated a yield of 50 Mt, or the power of 3,800 Hiroshima bombs detonated separately.[12]

The 8-kilometer-wide fireball reached nearly as high as the altitude of the release plane and was visible at almost 1,000 km (260 mi) away from where it ascended. The mushroom cloud was about 60 km (40 mi) high, which meant that the cloud was above the stratosphere and well inside the mesosphere when it peaked. The cap of the mushroom cloud had a peak of 95 km (59 mi) high.

All buildings in the village of Sverny, both wooden and brick, located 55 km (34 mi) from ground zero within the Sukhoy Nos test range, were destroyed. In districts hundreds of kilometers from ground zero, wooden houses were destroyed, stone ones lost their roofs, windows and doors were broken, and radio communications were interrupted for almost one hour. One participant in the test saw a bright flash through dark goggles and felt the effects of a thermal pulse even at a distance of 270 km (170mi). A shock wave was observed in the air at Dikson settlement 700 km (430 mi) away; windowpanes were partially destroyed up to 900 kilometers (560 mi) away.[13] Although the Tsar Bomba was detonated 4 kilometers above ground, a seismic shock wave equivalent to an earthquake of over 5.0 on the Richter-Scale was measured around the world.[14]

The Tsar Bomba, although the most powerful, was only one of more than 450 nuclear tests conducted by the Soviet Union between 1949 and 1981 at the Semipalatinsk test site (STS) in Kazakhstan. Little information is available on the health effects of exposure to fallout from Soviet nuclear weapons testing and on the combined external and internal environmental exposures that have resulted from these tests. A total of more than 10,000 people who inhabited areas northeast and southeast of the STS were considerably affected by fallout mainly from 118 atmospheric and surface nuclear tests carried out from 1949 to 1965. Local health authorities as well as international agencies have expressed concern with regard to radiation-associated health effects, in particular cancer.[15]

While the damage done in the STS over the years was serious, it is on the fragile ecosystems of the atolls and remote islands in the Pacific where even more damage has been done by nuclear bomb testing and little responsibility accepted by the countries that detonated these bombs. To this, we will now turn our attention.

Nuclear bomb testing by the United States in the Pacific began just two years after the end of the Second World War and continued intermittently until late 1962, running up a total of 105 tests of fission (A bomb) and fusion (H bomb) weapons, all of them in the atmosphere or in shallow lagoon waters. The tests were conducted not only to evaluate successive bomb designs and to attempt to improve on them, but also to observe the effects of nuclear explosions on ships and crews.[16]

Islanders were evacuated from Bikini atoll (now Pikini) successively to a series of other atolls, where they fared poorly because of unsuitable environments for fishing and growing crops. Their home islands on Eniwetak are still not safe. On Eniweke (now Enewetak) A-bomb tests began in 1948, followed by H-bomb tests in 1952, which later alternated between there and Bikini. A further twenty-four US tests were made as air drops close to Christmas atoll, a British island in the Line Islands chain. Two very high-altitude rocket tests over Johnston atoll in 1962 closed out the American atoll testing.[17]

Meanwhile the British at Christmas and Malden exploded nine devices and since 1966 the French at Mururoa and Fangataufa conducted 44 tests in the atmosphere and 115 underground tests in lagoon boreholes, finally ending all testing in 1996.[18]

Most noteworthy of these tests was the Castle series detonated by the Americans including the Bravo hydrogen test. The nuclear weapons testing program in the Pacific permanently altered the lives and future of the people of Rongelap, Rongerik and Ailinginae atolls. The Rongelap people experienced personal health injuries and involuntary displacement when fallout from the largest thermonuclear weapon ever tested by the United States forced their physical evacuation from 1954 and 1957. They lost access to a formerly viable healthy ecosystem and lost their ability and rights to safely live in their environment when they resettled on Rongelap (1957 to 1985). They became exiles (1985 to present) when they were finally provided with evidence of hazardous levels of radioactive contamination throughout their homeland.[19]

The novel feature of Bravo was that about 40 percent of the fuel was enriched lithium—6 deutride. The case of the bomb was made of aluminum rather than heavy steel. The unexpectedly high yield of 15 megatons was a surprise and resulted from the supposedly inert lithium—7, which, when struck by high-energy neutrons, could fragment into a helium and a tritium atom. The tritium caused very energetic fusion that increased the expected 4-8 megaton yield.[20] Bravo created a crater on the reef rim and adjacent part of the lagoon about 2 km wide and 75 m deep, and threw all the excavated debris into the sky. The cloud rose in only six minutes to a height of 40 km. After 8 minutes even the cloud bottom rose to about 15 km.

It took about two days for the military to send evacuation ships to the affected atolls of Rongelap, Rongerik, Utirik and Ailiginae, where people received estimated doses of radiation from 14 to 175 rd. Bikini atoll was eventually declared too contaminated for human habitation.[21] The Rongelapese continued to eat food and drink water tainted from radioactivity for the days between the Bravo test and their evacuation to Kwajelein. Both the food and water supplies were obviously contaminated with radioactive fallout but the Rongelapse had no other food and water options available to them. After the evacuation of the Rongelapese, the US government verified the contamination levels of the food and water that the Rongelapese had consumed before their removal. Four days after the Bravo test, the US government collected water samples on Rongelap that were found to be radioactive at a rate of two to twenty-five times above the AEC operational tolerances.[22]

The Rongelapese suffered non-lethal levels of radiation exposure as well as what they ingested in the food and water they consumed immediately after the Bravo test. Fallout particles stuck to the coconut oil that the people used in their hair. Children played with the fallout ash. Several hours after the Bravo test the people began to feel nauseated. Many vomited and exhibited flulike symptoms. By evening, the people's skin had begun to blister from contact with the fallout ash. After the people were evacuated, radiation lingered in the air and continued to settle on the atolls.[23]

The release of energy equivalent to thousands and millions of tons of TNT, and quantities of radioactive materials must be inherently accompanied by some degree of risk.[24] Exposures fifty years ago still have health implications today that will continue into the future. Prior to 1950, only limited consideration was given to the health impacts of worldwide dispersion of radioactivity from nuclear testing. But in the following decade, humanity began to significantly change the global radiation environment by testing nuclear weapons. By the early 1960s, there was no place on earth where the signature of atmospheric nuclear testing could not be found in soil, water and even in polar ice.[25]

Let us now consider the blast set off by a nuclear device. The blast effects are limited to areas near the site of the detonation. Partial damage to structures may extend for a nominal-sized bomb (equivalent in energy to 20,000 tons of TNT) to about two miles, and to ten miles and more for a high-yield weapon.

Significant amounts of heat radiation may be received up to two miles from the blast point of a nominal bomb and about twenty miles or more for a high-yield weapon. At the time of a nuclear detonation the light produced from a nominal bomb may temporarily blind a person if he or she looks directly at it, even from a distance of thirty to forty miles.[26]

External exposure is an important feature of nuclear risks from weapons testing. At the time of detonation, the almost instantaneous gamma rays released will be of concern up to about a mile for a nominal bomb and to a few miles or more for high-yield weapons. However, significant amounts of radioactive fallout may occur at more than 300 miles downwind from high-yield weapons.

About 25 roentgens of gamma radiation, delivered in a short time (about a day or less) are required to produce minor and transitory changes in the blood; about 100 roentgens are required for some persons to show radioactive sickness; about 450 roentgens may be lethal to half of the exposed persons. A roentgen is a unit for measuring the amount of radiation, or dose, which has been received. For example, a normal x-ray will deliver about one tenth roentgen or more to the chest; about 10 roentgens are received in a lifetime from cosmic rays and from naturally occurring radioactivity in the air, water, and soil.

Following the March 1, 1954, thermonuclear detonation at the Pacific proving ground known as the Bravo thermonuclear blast, a shift of the winds caused a heavy fallout over some Marshall Islands. The highest radiation exposure was 175 roentgens. Most of the people exposed experienced radiation sickness, but there were no deaths.[27] The now infamous press release by the Atomic Energy Commission (AEC) issued ten days after the Bravo test stated the following: "During the course of a routine atomic test in the Marshall Islands . . . individuals were unexpectedly exposed to some radioactivity. There were no burns. All were reported well. After the completion of the atomic test, the natives will be returned to their homes." The AEC knew at the time this statement was read that many Marshallese had suffered burns, nausea, and vomiting—symptoms of acute radiation sickness. The 15-17 megaton Bravo test was anything but routine.[28]

Fallout material also emits beta rays that travel at most a few meters in the air and are of principal concern when highly active fallout material remains in contact with the skin for relatively long periods of time. This may result in skin damage that would appear like a burn and may ulcerate if the radiation exposure has been long enough. Beta burns occurred on the skin of the Marshallese exposed to the fallout from the Bravo test.

Another facet of the effects of radiation on humans is internal exposure. Radioactive iodine present in fallout material may find its way into the body through ingestion or inhalation and will concentrate in the thyroid gland.

Because of their eating habits, grazing animals ingest materials over a relatively large area, which means that significantly larger amounts of radioactive iodine reach the thyroid. Experimental studies with sheep, for example, show that about 16,000 roentgens are required to produce even minimal changes in the cell structure and about 50,000 roentgens are required to produce definite cell damage and hypothyroidism.

Owing to the more widespread distribution of fallout from the Pacific tests in the spring of 1954, radioactive iodine was found in the thyroids of cattle and sheep in various parts of the United States. The highest total radiation dose to the thyroid was estimated to be about 40 roentgens.

One of the biologically important elements in fallout is strontium 90.[29] If it is taken into the body, it is selectively deposited in the bones and continues to irradiate the surrounding cells for long periods of time, since it has a half-life of about 27.7 years. The deposition of relatively large amounts of strontium—90 in the bone would be expected eventually to produce bone tumors. Owing to its relatively long half-life the amount of strontium-90 will accumulate in the environment if more is continually added.

Particles of neutrons that are released from a nuclear detonation react with a nitrogen atom in the air to form radioactive carbon, called carbon—14, which has a long half-life of about 5,600 years. This radioactive carbon enters into the biosphere, or the environment of living things, alongside normal carbon and this in turn finds its way into all living tissues and irradiates the surrounding cells. Since carbon is found in all living matter, the effects are similar to those when the whole body is irradiated by an external source.

Air and water usually contain more than one naturally occurring radioactive substance (radon, thoren, radium, potassium). The radioactive content in water supplies varies greatly, but some has been found in essentially all sources.[30]

Radioactive materials have been found in plankton and algae in higher concentrations than in the surrounding water. When fish eat these marine organisms, a small fraction of this radioactivity is retained within the fish. Fish containing a greater than the maximum possible amounts of radioactivity have been found at the test atolls of Bikini and Enwetok, where fish is used as a principal source of food, and in the nearby areas of the heaviest fallout. The presence of radioactivity in tuna fish was reported by the Japanese following the Bravo test in the Pacific.

Radiation can cause irreversible changes or genetic mutations in the germ cells. Most of these mutations are considered harmful. The number of mutations produced is independent of the rate of radiation exposure. That is, it is the total exposure to a given kind of radiation that is the important factor.

An important feature of nuclear tests and their resulting impact on humans and the environment is that this was done without open dialogue or discussion with those directly involved. Despite the British, Soviet and American agreement in 1963 to ban testing in the atmosphere, under water and in space, General de Gaulle of France had no problems about continuing atmospheric tests. Nor was he persuaded to halt testing by the continued protests by the elected representatives of the Polynesian people, who argued that detonating atomic bombs in the midst of 140,000 people could create serious health problems. When the local parliament, the Territorial Assembly of the Pacific atolls of Moruroa and Fangataufa, objected to the continuation of atmospheric tests, they were simply told by the French governor that since they lived in a colony, all questions relating to defense matters were outside of their competence.[31]

On July 2, 1966, the French tried out their new atomic test site at the Moruroa atoll. The first bomb, a plutonium fission device, was placed on a barge anchored in the lagoon. When it was detonated all the water in the shallow lagoon basin was sucked up into the air, and then rained down. The islets on the encircling reef were completely covered with piles of irradiated fish and clams, whose slowly rotting flesh continued to stink for weeks.[32]

On July 19, the French dropped the next bomb from an airplane flying 45,000 feet above the atoll above the ocean, 60 miles south of the atoll. Two days later, an untriggered bomb on the ground was exposed to a "security test." While it did not explode, the bomb's case cracked and its plutonium contents spilled over the reef. The contaminated area was "sealed" by covering it with a layer of asphalt.[33] These first two tests were a prelude to the bigger one later in September detonated in the presence of General de Gaulle himself on a warship equipped with protective iron shields and sprinklers for washing away radioactive dust. The "bomb," a box-like nuclear charge, was exploded despite unsuitable weather conditions. Monitoring stations set up by the New Zealand National Radiation Laboratory in the Cook Islands, Niue, Samoa, Tonga, Fiji and Tuvalu to the west of French Polynesia immediately registered heavy radioactive fallout.[34]

During the next eight years, another forty-four French bombs, including five hydrogen bombs, were detonated in the Pacific skies above Moruroa and Fangataufa, another small atoll 40 kilometers further south. The monitoring stations New Zealand operated on other Polynesian islands regularly registered heavy fallout. Vociferous opposition continued to grow in Australia, New Zealand and the other Pacific islands. In 1973, widespread boycotts of French goods, airlines and shipping lines began. That year Australia and New Zealand initiated proceedings against France in the international Court of Justice in The Hague. The tests were then moved underground.

An accident on July 25, 1979, occurred in which an estimated one million cubic meters of material was ripped out of the flank of the atoll. The full extent of the resulting leakage of radionuclides into the ocean due to this accident is unknown. In 1981, two typhoons hit Moruroa. Civilian technicians released a secret report to the French press, revealing that this storm had washed out to sea the huge amounts of nuclear waste that had been allowed to accumulate on the north coast. This waste included 10-20 kilograms of plutonium that had been spilled out on the reef between 1966 and 1974 during the "security tests" and later covered by asphalt. The plutonium was scattered over the lagoon.[35] This plutonium waste dispersed by the typhoons presented a great danger to the health of the Pacific islanders according to the World Health Organization. None of these potential health problems have ever been addressed.

All the inhabitants of French Polynesia, who now number 188,000, face the insidious hazard of the steady absorption of radioactive fallout resulting from the forty-four nuclear tests in the atmosphere between 1966 and 1974. As surveys made in Micronesia show, it takes between ten and fifteen years before the effects of fallout become apparent. The most common radiation induced diseases are leukemia, brain tumors, and thyroid cancers. As could be expected it is from the 1980s that a sharp increase in the number of these three types of cancer has occurred in French Polynesia.[36]

In 1954, the Castle nuclear test series started in the Pacific. Bravo was fired on March 1. The Bravo device weighed only twelve tons. It was expected to yield six megatons, but in fact gave a yield of fifteen megatons, a thousand Hiroshimas and more than twice what was expected.[37] The Daigo Fukuryo Maru No. 5, a 99.9 ton fishing vessel with 23 crewmen on hoard and carrying the incorrect English translation of "Lucky Dragon," was catching fish outside of the danger zone of the blast, which the US government had declared in advance of the morning of March 1. The fishermen observed a brilliant, reddish light to the west south-west of their vessel. Two blasts were heard before a mushroom-shaped cloud appeared on the horizon. The cloud expanded, eventually covering the sky with darkness.

On first seeing the light, the fishermen began hauling in their lines. About three hours later, white ash began to fall on the deck where they were working. The crew processed the catch (mainly tuna) on the upper deck. By noon, the decks were so covered that the fishermen were making "visible footprints" in the white dust. Within two to three days, all of them became ill, suffering from headaches and nausea. A week later the crew experienced acute radiation syndrome (ARS) consisting in painful irritations, from burns to skin exposed to the ash—on the neck, the face, the ears and the area where they wore the *haichi maki* which is a cotton towel wrapped around the head. On March 14, the fishermen arrived back at their home port of Yaizu.[38] During their ARS treatment back in Japan, the crew were inadvertently infected with hepatitis through blood transfusions. All recovered except for Aikichi Kubyoma, who died on September 23, 1954, from an underlying liver cirrhosis compounded by the secondary hepatitis infection. Kuboyama is considered the first victim of the hydrogen bomb and of the Castle Bravo test shot.[39]

But it was not the Japanese fishermen but rather the Rongelap people themselves who suffered the most from the Bravo nuclear bomb test. The hydrogen test-bomb Bravo dropped on the Marshall Islands in 1954 had enormous consequences for the Rongelap people.[40] The Rongelapese residing on Rongelap and Ailinginae were exposed to what was then estimated to be 200 roentgens of whole-body radiation, which is recognized as a potentially lethal dose, with substantially greater amounts taken up by the thyroid. Children were exposed to even greater levels of radiation.[41] Anthropologists Barbara Rose Johnston and Holly Barker provide incontrovertible evidence of physical and financial damages to individuals and cultural and psycho-social damages to the community.

The US nuclear testing program was conducted in the Marshall Islands from 1946 through 1958. The US government detonated atomic and thermonuclear weapons with the aim of achieving world peace through a deterrence policy. The US nuclear testing program essentially inflicted nuclear war conditions on a fragile atoll ecosystem and on a vulnerable population. The Marshallese, despite appeals to the United Nations, were powerless to stop the testing and unprepared to address the proliferation problems resulting from the testing.[42]

Human environmental impacts of nuclear testing included contamination and exile as experienced by the people of Rongelap, Aslinginae, and Rongerik atolls. The nuclear weapons testing destroyed the physical means to sustain and reproduce a self-sufficient way of life for the people of Rongelap. This resulted in contamination, short and long-term exposure to radioactive substances, and alienation from land and other critical resources. Radioactive contamination and involuntary relocation radically altered health, subsistence strategies, sociopolitical organization and community integrity.[43] A lifetime of service as unwilling human subjects in a wide variety of biomedical experiments further harmed the health and psychosocial well-being of the people of Rongelap.

The people of Rongelap experienced involuntary displacement from their homes on Rongelap and Ailinginae atolls when they were physically removed from the atolls (March–May 1946; 1954–1957). When these people returned to their atolls, they lost access to a viable and healthy ecosystem. This deprived them from the ability and their rights to safely live in their environment in the years between 1957 and 1985. They then became exiles (1985-present) when they were finally informed of the life-threatening contamination in their homeland

Exposure concerns involved much more than exposure to radiation and fallout from the Bravo test in 1954. The worries involve the persistent presence of contamination from the sixty-seven atmospheric tests of nuclear weapons in the Marshall Islands. This contamination includes radioactive elements released through nuclear explosions as well as tracer chemicals such as arsenic used to "fingerprint" the fallout from each weapon.

In addition to external radiation the people of Rongelap, Rongerik, and Ailinginae were exposed to internal radiation such as breathing dust and smoke from household and garden fires, drinking contaminated water, and consuming terrestrial and marine food sources, as well as living in houses and using materials fashioned from contaminated materials. Future generations are also at risk due to the synergistic, cumulative and genetic effects of long-term exposure and other contamination from nuclear testing. New taboos appear: certain lands and foods are off-limits; marriage to certain people involved new social stigmas; and birthing presented new fears and health risks. Family life often involved the psychological, social and economic burden of caring for the chronically ill and disabled.[44]

One of the main concerns of the Marshallese in the 1980s was the contamination of their islands. Fearful of their continued exposure to radiation the Rongelapsee decided to abandon their island in 1985. Assisted by Greenpeace, they were resettled on the tiny island of Majetto, located on the remote north-west quadrant of Kwajalein Atoll. Once there, the Rongelapese discovered that the land was so sub-fertile that they had to depend on US Direct Aid ships for food. Due to these unsatisfactory arrangements, the Department of Energy later informed them that if they followed certain dietary and behavioral guidelines, they could safely return to Rongelap. Many difficult questions remained unanswered, however. "How much local food can we eat? Should our children be allowed to play in the sand? How many times a day should we wash our hands? How do we keep the hands of our children out of their mouths? Is it safe for a baby to drink its mother's milk?"[45] None of these questions have yet received adequate answers.

China also carried out a vigorous nuclear testing regime. The Lop Nor Nuclear Weapons Test Base is located in the Malan, Xinjiang Autonomous Region of China where China conducted forty-five nuclear tests, including twenty-two atmospheric and surface tests between 1964 and 1981. The test site is the world's largest, occupying an area of 100,000 sq. km. Lop Nor contains four testing zones, three used for underground testing and one used for atmospheric tests. Only two of the zones are currently used, which occupy an area of about 200 sq. km.

The people residing in settlements in Makanchi, Urdzhar and Tashesken are located near the Kazakhstan-Chinese border. They were exposed to ground and atmospheric nuclear tests at Lop Nor and are believed to have been heavily exposed to radioactive fallout.[46]

The United Kingdom also engaged itself in nuclear weapons testing. 20,000 men from the UK took part in the UK's programs of atmospheric tests in Australia and the Pacific Ocean during the 1950s and 1960s. Between 1952 and 1958 the Ministry of Supply conducted a series of atmospheric nuclear weapons tests in Australia and its islands in the Pacific Ocean. Other experiments in which radioactive materials were dispersed into the environment were also carried out by the Ministry of Supply in South Australia between 1953 and 1963. Survey and clean-up operations continued until 1967. Staff from the United Kingdom also participated in American tests at Christmas Island in 1962 and finally vacated the island in 1964.

Results suggest that participation in nuclear tests was associated with an increased mortality from leukemia, multiple myeloma, and other injuries and poisoning but, surprisingly, a decreased mortality from cancers of the prostate and kidney and from chronic bronchitis.[47]

India and Pakistan have also conducted nuclear tests. The series of underground nuclear tests conducted by India and Pakistan in May 1998 has substantially altered the regional and global security environment. On May 11 and 13, 1998, India set off five nuclear devices of which one was a hydrogen bomb at its test site in Pokhran in the northwestern Indian state of Rajasthan—the first such tests in twenty four years. The initial test had been carried out at the same site on May 18, 1974 and was billed as a "peaceful nuclear explosion." This first test had a 15-kiloton yield.[48]

A central assumption behind the American nuclear testing policies in the 1950s was that US national security was more important than the welfare of the peoples living and working near the proving grounds.[49] The Atomic Energy Commission (AEC) of the United States faced a conflict of interest over the issue of nuclear testing. Fallout from both the atomic tests and H-bomb explosions in Nevada and elsewhere through the 1950s and the H-bomb explosions raised a new and serious threat to public health. The continental tests produced local fallout that killed sheep on western ranges and exposed citizens in communities such as St. George, Utah, to serious levels of radiation. Yet the AEC intent on continuing the Nevada tests, which were far cheaper and more convenient than those in the distant Pacific, kept a tight lid on the relevant data while insisting that there was no public danger.

This created a clear conflict in having the AEC responsible for conducting the weapons tests and for providing the administration with the facts about fallout. Desire for perfecting new and more effective weapons took precedence over public disclosure.[50]

NOTES

1. Von Hippel, Frank N. 2019. "The Decision to End US Nuclear Testing," *Arms Control Today*, Dec. 2019, *Arms Control Association*, http://www.armscontrol.org/act/2019-12/features/decision-end-us-nuclear-testing visited Dec. 2020.

2. Simon, S.L, Bouville. A. 2002. "Radiation doses to local populations near nuclear weapons test sites worldwide," *Health Physics* 82 (5), pp. 706-725.

3. Brown, Kate. 2017. *Manual for Survival. A Chernobyl Guide to the Future*. New York: Norton, p. 311.

4. Ibid.

5. Simon, Steven L. 2015. "Health Effects of nuclear weapons testing," *The Lancet*, August 1-7, Vol. 386 (9992), p. 407.
6. Ibid.
7. National Research Council. 1999. "Exposure of American people to Iodine-131 from Nevada nuclear bomb tests," Review of the National Cancer Institute Report and Public Health Implications 1999, National Research Council, Washington, DC.
8. Brown, p. 311.
9. The Legacy of Nuclear Testing, ICAN, www.icanw.org/thefacts/catastrophic-harm/the-legacy-of-nuclear-testing.
10. Brown, *Manual*, p. 311.
11. Ibid.
12. Tsar Bomb, Atomic Heritage Foundation. 2014. atomicheritage.org/history/tsar-bomba. August 8.
13. Big Ivan, The Tsar Bomba, 2007. The Nuclear Weapon Archive. A Guide to Nuclear Weapons, 31 August, nuclearweaponarchive.org/Russia/TsarBomba.html.
14. Comprehensive Nuclear Test-Ban Treaty Organization, 1961. The Tsar Bomba The Most Powerful Man-Made Explosion Ever, CTBTO Preparatory Commission, ctbto.org/specials/testingtimes/30-october-1961-the-tsar-bomba.
15. Bauer, Susanne, Gusev, Boria I. Pivina, Ludmilla M., Apsalikov, Kazbek N., 2005. "Radiation exposure due to local fallout from Soviet atmospheric nuclear weapons testing in Kazakhstan; solid cancer morbidity in the Semipalantinsk historical cohort, 1960-1999," *Radiation Research*, October 2005, Vol. 164 (4 Pt 1) pp. 409-419.
16. Winterer, Edward L., "Nuclear Bomb Testing," in *Encyclopedia of Islands*, Ed. Gillespie, Rosemary G. and Claque, David A. 2009. Berkeley, CA: University of California Press, p. 680.
17. Ibid.
18. Danielsson, Bengt. 1990. "Poisoned Pacific: The Legacy of French Nuclear Testing," *Bulletin of the Atomic Scientists V.* 46, No. 2 (March 1990), p. 22.
19. Johnston, Barbara Rose and Barker Holly Rose. 2008. *The Consequential Damage of Nuclear War. The Rongelap Report*, Walnut Creek: Left Coast Press, p.89
20. Reed, Thomas C. 2004. *At the Abyss. An Insider's History of the Cold War*. New York: Ballantine Books, 2004, p. 78.
21. Johnston and Barker, p. 96.
22. Joint Task Force Seven, 1956. *Radioactive Survey of Downwind Atolls Contaminated by Bravo*. Washington, DC: Joint Task Force Seven.
23. Johnston and Barker, p. 98.
24. Dunning, Gordon M. 1955. "The Effects of Nuclear Weapons Testing," *The Scientific Monthly*, Vol. 81, No. 6 (December), p. 265., New York Operations Office, Health and Safety Laboratory, Jan. 21, 1955, 7, http: worf.eh.doe.gov/data/ihpid/39683e.pdf
25. Simon, Steven L., Bouville, Andre, Land, Charles E. 2006. "Fallout from Nuclear Weapons Tests and Cancer Risks," *American Scientist*, V. 94, N. 1, January, 48.
26. Ibid, p. 266.
27. Libby, Willard F. Murray, Thomas E. von Neumann, John Strauss, Lewis.1955. "Radioactive Debris from Operation Castle: Worldwide Fallout," New York: United States Atomic Energy Commission, OSTI. GOV. pp. 77-92.
28. Johnston and Barker, p. 45.
29. Ibid, p. 267.
30. Ibid, p. 268.
31. Danielson, Poisoned Pacific, p. 24.
32. Ibid.
33. Ibid, p. 25.
34. Ibid.
35. Ibid, p. 27.
36. Ibid, p. 28.
37. Reed, Thomas C. *At the Abyss*, p. 78.

38. Divine, Robert. 1978. *Blowing on the Wind: The Nuclear Test Ban Debate 1954-60.* New York: Oxford University Press, Background Chapter, p. 6.
39. Ibid
40. Johnston and Barker, p. 43.
41. Ibid, p. 96.
42. Ibid.
43. Ibid, p. 44.
44. Ibid, pp. 45-6.
45. Testitimony by Billiet Edward, 1994. Oversight Hearing before the Subcommittee on Natural Resources, House of Representatives, 103rd Cong., on "Radiation Exposure from Nuclear Tests in the Pacific," Serial No. 103-68, 24 Feb. 1994, Washington, DC, p. 382.
46. Zhumadilov, Kassyn, Ivannikov, Alexander, Zharlyganova, Dinara, Stepanenko, Valery, Zhaxybay, Apsalikov, Toyoda, Shin Endo, Atoru, Tanaka, Miyazawa, Tanaka, Titsuji, Okamoto, Tanaka, Tetsuji Chouzou, Masaharu Tetsuji Chouzou, Hoshi. 2011. "The influence of the Lop Nor Nuclear Weapons Test Base on the population of the Republic of Kazakhstan," *Radiation Measurements*, Vol. 46(4), pp. 425-429.
47. Darby, S. C., Kendall, G. M, Fell T. P., et al. 1988. "A summary of mortality and incidence of cancer in men from the United Kingdom who participated in the United Kingdom's atmospheric nuclear weapons tests and experimental programs," *British Medical Journal*, 296, Issue 6618, p. 336.
48. Thomas, Raju G. C., and Gupta, Amit. 2000. "Introduction," *India's Nuclear Security*, Edited by Thomas, Raju G. C., and Gupta, Amit. Boulder, London: Lynne Rienner Publishers. 2000, p. 1; Ganguly, Sumit. 1999. "India's pathway to Pokhran II," *International Security*. Vol. 23:4, Spring, p. 148.
49. Smith-Norris, Martha. 1997. "Only as Dust in the Face of the Wind. An Analysis of the BRAVO Nuclear Incident in the Pacific, 1954," *Journal of American- East Asian Relations*, Vol. 6, No. 1 (Spring), p. 34.
50. Devine, R.A. 1989. "Atoms for Peace and War, 1953-1961," *Review Science* 10 Nov. Vol. 246 (4931), 826 pdf.

ELEVEN

Pakistan

The Biggest and Most Important Exporter of Nuclear Technology in the History of the Nuclear Age

Earlier in the book the risk of a nuclear war between India and Pakistan was the focus of our attention and during the discussion the importance of A. Q. Khan in the development of the nuclear bomb in Pakistan was discussed as well as the part he and his network played in this proliferation of nuclear technology. We will now return again to the activities of Khan and his proliferation as well as that of Pakistan as a central topic in the present chapter. Its main thesis is that Pakistan is a most important exporter of nuclear technology.

To understand the work of A. Q. Khan on behalf of Pakistan and the way that nuclear non-proliferation was compromised by this activity it is necessary to understand the difference between Uranium 235 and Uranium 238 and the way that the former is enriched from the latter by using centrifuge technology.

The uranium atom is the heaviest atom present in the natural environment. Uranium is the only natural nucleus that easily undergoes fission. Highly sought after, it can be used as a fuel in nuclear reactors but also as an explosive in atomic bombs.

Uranium 238 is more abundant than 235. What can happen to 238 is that a neutron capture causes the nucleus to become unstable. After a few days, Uranium 238 has transformed itself into fissile plutonium PU 239, formed by neutron capture from U-238 and which when fissioned yields much the same energy as the fission of U-235. PU 239 is a radioisotope with a half-life of 24,000 years. Plutonium is highly fissile and can also be used as a nuclear fuel or an atomic explosive. Until the arrival of A. Q.

Khan in the nuclear weapons development program in Pakistan, the effort was to develop Plutonium as a fuel for their atomic weapons. While this work continued, he persuaded the authorities to allow him to work with Uranium 235.[1]

The commercial process used by him for the enrichment of uranium in the U-235 isotope involved gaseous uranium in centrifuges. Before enrichment, uranium oxide had to be converted to a fluoride so that it could be processed as a gas, at a low temperature.

From a non-proliferation standpoint, uranium enrichment is a sensitive technology which needs to be subject to tight controls. It is these tight controls that A. Q. Khan chose to violate.

Natural uranium contains 0.7 percent of the U-235 isotope. The remaining 99.3 percent is mostly the U-238 isotope which does not contribute directly to the fission process. Isotope separation is a physical process to concentrate ("enrich") one isotope relative to others.

Uranium-235 and U-238 are chemically identical but differ in their physical properties, most notably in their mass. The nucleus of the U-235 atom contains 92 protons and 143 neutrons, giving it an atomic mass of 235 units. The U-238 nucleus also has 92 protons but has 146 neutrons—three more than U-235—and it therefore has a mass of 238 units.

The difference in mass between U-235 and U-238 allows the isotopes to be separated and makes it possible to "enrich" the percentage of U-235. All present and historic enrichment processes directly or indirectly make use of this small difference in mass.[2] The nuclear black market run by Khan and his Pakistani overseers was based on this small difference.

Just after midnight on October 4, 2003, the German-owned freighter *BBC China* made an unscheduled stop for only two hours. The German Secret Service had contacted the ship's owners in Hamburg requesting their help. As it passed through the Suez Canal the owners instructed the crew to make a change in course. "Don't ask any questions" the Captain was told. At the dock in Taranto, Italy, a team made up of America's CIA and Britain's Intelligence Service (formerly known as MI6) realized that they would have to work quickly to identify five forty foot cargo containers and then remove them from among more than two hundred other containers on board. The *BBC China* then had to recommence its voyage, the ultimate destination of which was the Libyan capital Tripoli.

The five containers' journey had begun in August in a factory in Malaysia. A CIA and Intelligence Service team had received a tip-off in mid-September that a consignment of important goods had arrived at Dubai's free trade zone where they were loaded onto the *BBC China.* The ship was closely monitored. But until the five containers were opened the team could not be certain that their suspicions were correct. The opening of the containers led the operatives to breath sighs of relief. Inside were thousands of aluminum components packed in wooden boxes marked with the logo SCOPE. Only a trained expert would have known that these were intended for a centrifuge, the device that enriches uranium, the heart of the clandestine nuclear program that Khan and Pakistan were promoting.

Within hours, a senior MI6 officer would call on his Libyan counterpart and demand an explanation. Soon after the Libyan leader would surprise the world by announcing that he was giving up something that few knew he had, namely an active nuclear weapons program.

As for the man who had provided the contents of the five containers on the *BBC China*, this was also the beginning of the end. As well, the end was slowly beginning on the decades long career of Pakistani scientist, spy, and national hero A. Q. Khan, the man who in the 1970s was responsible for building Pakistan's bomb, and who had long been operating as a salesman of the most dangerous and most closely guarded nuclear secrets. A week before the *BBC China* interception, the CIA director, George Tennet, had begun the process of putting Khan out of business in a meeting with a stunned President Musharraf of Pakistan. What stunned Pakistan's President was not the news of the activities of A. Q. Khan but just how much the US government knew about them. Within hours of the *BBC China's* interception a senior US diplomat would arrive in Islamabad to insist that the time for action had come. The end had begun for Khan and his network. But Khan and his network had already accomplished untold damage by spreading secret and dangerous technology across the world.[3] The 2003 disclosures about the clandestine sale of Pakistan's nuclear technology not only to Libya but also to Iran and North Korea forced the authorities in Islamabad to move against Khan[4] notwithstanding the fact that they had co-operated in his nefarious activities of exporting nuclear technology to other countries. As MIT scholar Jim Walsh put it: "Pakistan is absolutely the biggest and most important exporter of nuclear technology in the history of the nuclear age."[5] Thanks to A. Q. Khan, for the first time in history all the elements of nuclear weapons development, the supplier networks, the material, the centrifuge and enrichment technology, and the warhead designs—were outside direct state control at least for part of the time during this roughly sixteen year proliferation crisis.[6] They were outside of Pakistan's direct control, but not outside its indirect control.

The Honorable Gary J. Ackerman, US Congress Chairman of the Subcommittee on Terrorism, Non-Proliferation and Trade for the Middle East and South Asia pointed out that A. Q. Khan, the father of Pakistan's nuclear weapons program, for many years conceived and operated an illegal international nuclear proliferation network. He sold nuclear equipment to North Korea and Iran and sold the same equipment and technologies along with weapons' design to Libya.

> To Iran, Khan provided centrifuges, technical designs, components and an address book of suppliers for other material. We do not know

> whether he provided Iran with the same plans that he provided to Libya. But what we do know is that Iran's nuclear program is years ahead of where it would have been without his assistance.
>
> To North Korea Khan supplied centrifuge machines along with drawings, technical data, depleted uranium hexafluoride and a shopping list so that North Korea also could produce additional equipment directly from foreign suppliers. In return Pakistan got missiles.
>
> To Libya, Khan committed to supplying the entire enrichment process and the design for a nuclear weapon. In return, Khan got wealthy.
>
> What is most startling is not the scope of Khan's network that encompassed at least ten countries and involved at least thirty middlemen, but that so few countries, companies, or individuals have so far been held accountable. Apparently the stiffest penalty the Pakistani Government can impose on those who sell nuclear secrets is house arrest. Elsewhere around the world only a handful of Khan's co-conspirators have faced criminal charges. Most continue their business lives without let or hindrance and probably also continue their trade in nuclear-related materials.
>
> Pakistan still has a nuclear program that operates largely without international scrutiny or voluntary transparency. And because Pakistan is not a signatory to the Nuclear Non-Proliferation Program, it will still have to produce nuclear related materials and technology clandestinely in order to sustain the program.[7]

The deficit of scrutiny of Pakistan's nuclear weapons program was not only that of the international community but also that the elected politicians of Pakistan. From the time of Zulfikar Ali Bhutto's overthrow in 1977, the only people in authority definitely not involved in the nuclear program were elected politicians. Zulfikar Bhutto maintained control because it had been his idea to acquire atom bombs in the first place. His daughter Benazir Bhutto, who followed him as Prime Minister of Pakistan from 1988 to 1990 and from 1993 to 1996 had promised the Americans, according to General Hamid Gul, head of the all-powerful Inter-Services Intelligence Agency during her first term in elected office and a devoted friend of hardline jihadists, that she would end development of nuclear weapons. But when she took office, she found that this was not very easy to do.

According to Hassan Abbas, author of *Pakistan's Nuclear Bomb. A Story of Defiance, Deterrence and Deviance*, Bhutto told him that within weeks of her becoming prime-minister General Beg, who was then the chief of the army, emphasized to her that the nuclear program was not something she should know about and referred her on this matter to the president, Ghulam Ishaq Khan. When she asked the president about it, he too rebuffed her: "There is no need for you to know."[8]

In 1989 Benazir went to Washington and told a joint session of Congress there "I can declare that we do not possess, nor do we intend to make a nuclear device." The day before her speech she was given an unprecedented briefing by CIA director William Webster on what Americans knew about the Pakistani bomb. Bhutto was said to be shocked. But she was aware that her nuclear assurances would be hollow. Her army chief later admitted that she was "putting on a show for the Americans."[9]

The nuclear issue was decisive for the continuance of Benazir's first term as prime minister. She was up against a coalition of scientists, businessmen, military officers and civilian officials who saw the bomb not only as vital to the national interest but also, through the proliferation of nuclear technology, as a lucrative source of funds for the state—and for many of the individuals involved.[10] And therefore Ishaq Khan eventually removed her from power.[11]

Another Pakistani who was in favor of and participated in nuclear proliferation was Mirza Aslam Beg, the former Chief of Pakistan's army, who used the social media to make the following controversial encouragement to fellow Muslims: "All Muslim countries should have nuclear weapons to safeguard themselves. Any who don't get them are cowards in my opinion." Turning his attention to non-Muslims he tweeted: "Have talked to army chief of staff of North Korea and asked him not to disarm nuclear weapons because North Korea is the strongest force against American aggression and many countries including Pakistan learn and inspire (sic) from them." This was followed by: "I am hearing news that North Korea may disarm all its nuclear weapons. This is really sad news if true. Especially when we gave them the nuclear technology." This is one of the clearest statements yet from a military insider that Pakistan did indeed sell North Korea the bomb.[12]

There are three critical periods when important nuclear proliferation decisions were taken: 1987–1989 (Iran), 1993–1995 (North Korea and Iran), and 1997–1999 (North Korea and Libya). In the second stage (contact with North Korea from 1992), Khan represented the government of Pakistan possibly in league with the military leadership while the civilian leadership were left in the dark. During the third stage (beginning in 1997–1999) Khan started operating independently, and the Libyan deal was finalized during this period. He also continued to deal with North Korea on his own.[13] However, research on this subject falls short of fully explaining what led to this proliferation crisis.[14]

The search for the motivations behind Khan's choices is complex. All of his own arguments revolved around the notion that Pakistan was now capable of dealing with the Indian threat given that Khan and his associates had made the country's security foolproof through the nuclear option.[15]

Pervez Musharraf in his memoirs forcefully argued that "the show was completely and entirely A. Q.'s and he did it for the money."[16] Subsequent to his exposure Khan agreed to sign a twelve page confession, which has never been made public, but parts of it he repeated later on television.[17] Musharraf decided to pardon Khan in recognition of his previous contribution to Pakistan's acquisition of a nuclear capability, and even allowed him to retain the illegal money he had made in the process.[18] In 2008, Khan retracted his confession.

A. Q. Khan's nuclear proliferation activities were a combination of factors that included his own personal ambitions and his inward-looking worldview; the Pakistan army's need to compete with Indian missile development; weak state institutions; fragmented decision-making processes; and compartmentalized authority structures. The political chaos in Pakistan and the absence of civilian oversight and effective controls enabled the network to sell its nuclear merchandise and expertise throughout the world.[19] The Pakistani government was not directing this network. It was essentially a criminal operation, and therefore it was a more disturbing and dangerous operation than if it had been a secret government-controlled effort.[20] The IISS report "Nuclear Black Markets" argues that the official Pakistani line that the exports to Iran, North Korea and Libya were the work of one errant man and his duped associates cannot be taken at face value. Neither does the claim that Khan was merely a front man, doing the government's bidding in each of these cases appear to be valid. The report concludes by asserting that "a careful analysis shows that most of Khan's dealings were carried out on his own initiative."[21]

In July 2001 American intelligence agents tracked a Pakistani cargo aircraft as it landed at a North Korean airfield and took on a secret payload: ballistic missile parts, the chief export of North Korea's military. North Korea provided General Musharraf with the missile parts he needed to build a nuclear arsenal capable of reaching every strategic site in India. Pakistan was desperate to counter India's superior military force. The Pakistan military had long coveted the plans for North Korea's Nodong missile. By April 1998, it successfully tested a version of the Nodang, renamed the Gauri. Its flight range of about 1000 miles put much of India within reach of Pakistan's nuclear warheads. For its part, North Korea tried to replicate Pakistan's success in developing nuclear weapons based on uranium, one of the natural commodities that North Korea has in plentiful supply.

In return for the missile parts, Pakistan provided the North with many of the designs for gas centrifuges and much of the machinery it needed to make highly enriched uranium for the country's nuclear weapons project.[22]

Many dismiss the claims of Khan's confession and the Musharraf pardon. In an article in the *New Yorker* entitled "The Deal" Seymour Hersh quotes a senior intelligence official:[23]

> One thing we do know is that this was not a rogue operation. How do you get missiles from North Korea to Pakistan? Do you think A. Q.

> shipped all the centrifuges by Federal Express? The military has to be involved, at high levels.

The famous Pakistani physicist Professor Pervez Hoodbhoy similarly argues:[24]

> These centrifuges weighed something like half a ton each. You can't put them in your coat packet and walk away with them. . . . If there were aircraft of the Pakistan air force that flew these centrifuges out—well, obviously there had to be somebody at the top who was also involved.

Dr. Mohamed El Baradi, the Director-General of the International Atomic Energy Agency at the time called Khan's activities only "the tip of an iceberg" and claimed that "Dr. Khan was not working alone."[25]

The most disturbing aspect of the story of Khan's network is how poorly the nuclear non-proliferation regime succeeded in exposing and preventing the network's operation. Khan, with the help of associates on four continents, managed to buy and sell key nuclear weapons capabilities for more than two decades while eluding the world's best intelligence agencies and non-proliferation institutions and organizations. Despite a wide range of hints and leads, the United States failed to thwart the network throughout the 1980s and 1990s as it sold the equipment and expertise needed to produce nuclear weapons to major US enemies including Iran, Libya and North Korea.[26]

The Khan network succeeded for many years by exploiting weaknesses in export control systems and ways of recruiting suppliers, including some in states that were members of the Nuclear Suppliers Group (NSG). This is a multilateral export control regime and a group of nuclear supplier countries that seek to prevent nuclear proliferation by controlling the export of materials, equipment and technology that can be used to manufacture nuclear weapons. Khan's network had as its key customer states that were contemptuous of NSG controls and were committed to violating the Nuclear Non-Proliferation Treaty (NPT) in their quest for secret nuclear capabilities.

The Khan network is, first and foremost, an elaborate and highly successful illicit procurement program, that Khan created starting in the 1970s to supply Pakistan's gas-centrifuge program, which had been used to produce weapons-grade uranium for Pakistan's nuclear weapons program. Khan and his associates slowly expanded their import-export operation, however, into a transnational illegal network that also exported gas centrifuges and production capabilities, as well as designs for nuclear weapons to other, mostly Muslim, countries to make a profit and provide additional business for their international collaborators. In addition to money, Khan was also motivated by pan-Islamism and hostility to Western controls on nuclear technology.[27]

In April 2004, the United Nations Security Council passed Resolution 1540, which requires all states to criminalize proliferation to non-state actors and to establish, review, and maintain appropriate and effective export control systems. This resolution fills an important gap in existing nonproliferation regimes by including an export control law requirement for all 191 UN member states and targeting non-state actors.[28]

Another policy area undergoing change as a result of the revelations about the Khan network is the Proliferation Security Initiative (PSI) which the US Department of State defines as "a global effort that aims to stop shipments of weapons of mass destruction (WMD), their delivery systems, and related materials worldwide."[29]

In addition to the nuclear proliferation issue, there are also serious concerns about the security of the nuclear weapons in Pakistan itself. Pakistan's status as a nuclear power which was made manifest when it carried out six nuclear tests in May 1998, came under renewed international scrutiny following the terrorist attacks on the United States in September 2001. Of most immediate concern was the security of Pakistan's nuclear arsenal, which was believed to be vulnerable to Muslim terrorist groups with close links to the al-Qaeda organization, headed by Osama bin Laden.[30] Various extremist religious organizations operating in South Asia and the Middle East are suspected of having designs on the nuclear assets of Pakistan.

When Pakistan was developing its nuclear weapons infrastructure in the 1970s and 1980s its principal concern was that India would capture its nuclear weapons facilities if these facilities were located near the Pakistan-India border. For this reason, Pakistan has located much of its nuclear weapons infrastructure to the north and west of the country. As a result, most of Pakistan's nuclear sites are close to or even within areas dominated by Pakistani Taliban militants and the latest home of Al-Qaeda.

Recurrent political instability and a strengthening crime-terror nexus in Pakistan since 2007 have increased the vulnerability of the state's security infrastructure. A handful of these attacks targeted critical installations close to Pakistan's nuclear facilities. These include an attack on a nuclear missile storage facility at Sargodha on November 1, 2007, an attack on Pakistan's nuclear airbase at Kamra by a suicide bomber on December 10, 2007, the targeting of a bus carrying employees of the Pakistan Atomic Energy Commission (PAEC) on September 4, 2007, and, most significant of all, the August 21, 2008, attack when Pakistani Taliban suicide bombers blew up several entry points to one of the armament complexes at the Wah cantonment, considered one of Pakistan's main nuclear weapons assembly sites.[31]

The international crisis centering on the war against terrorism necessarily revives the debate about Pakistan's involvement in the nuclear non-proliferation regime and the need for tighter curbs on the export of nuclear technology to politically volatile regions in the Muslim world.

The single most important difference between Western and Pakistani perspectives on Pakistan's nuclear program and its emergence as a nuclear power is the question of Pakistan's stance on non-proliferation. The prevailing Western discourse on nuclear disarmament is deemed by Pakistan not to be responsive to its concerns about national security. In Pakistan the nuclear option has become an issue as much about national sovereignty as national security. Pakistan's nuclear policy is increasingly oriented to furthering Pakistan's regional ambitions. These debates are likely to be overshadowed by Pakistan's own perception of the danger it faces of a military threat from a nuclear armed India, its aspirations as a regional power; and its desire to be recognized as "First among equals" in the Muslim world community of nations. Any attempt to draw Pakistan into the non-proliferation regime and to persuade it to assume responsibilities as a nuclear power will need to address these issues. Only a concerted effort by all parties to involve Pakistan more closely in a global system of surveillance aimed at tightening controls on the export of sensitive material will really advance the cause of non-proliferation in South Asia.[32]

NOTES

1. Levy, Adrian, and Scott-Clark, Catherine. 2007. *Deception Pakistan, the United States and the Secret Trade in Nuclear Weapons,* New York: Walker & Company, pp. 34-35.
2. World Nuclear Association, world-nuclear.org/information-library/nuclear-fuel-cycle/conversion-enrichment-and-fabrication/uranium-enrichment.aspx visited November 2019.
3. Corera, Gordon. 2006. *Shopping for Bombs Nuclear Proliferation, Global Insecurity, and the Rise and Fall of the A. Q. Khan Network.* Oxford: University Press.
4. Abbas, Hassan. 2018. *Pakistan's Nuclear Bomb. A Story of Defiance, Deterrence and Deviance.* Oxford: University Press.
5. Pan, Esther. 2004. "Nonproliferation—The Pakistan Network," *Council on Foreign Relations* Vol. 83, No. 1, 12 February.
6. Oren Clary, Christopher. 2019. "The A. Q. Khan Network: Causes and Implications," Master's Theses, Naval Postgraduate School, Monterey, December, 2005 http://www.fas.org/irp/eprint/clary.pdf, visited November 2019.
7. A. Q. Khan's Nuclear Wal-Mart: Out of Business or Under New Management. 2007. Joint Hearing before the Sub-Committee on the Middle East and South Asia and the Subcommittee on Terrorism, Non-Proliferation and Trade, Committee on Foreign Affairs, House of Representatives, One Hundred Tenth Congress, First Session, 27 June 2007, Serial No. 110-97, p. 2.
8. Abbas, p. 187.
9. Macfarquhar, Emily. 2019. "Benazir and the Bomb," *The Alicia Patterson Foundation,* November 14, aliciapatterson.org/stories/Benazir-and-bomb visited November 2019.
10. Bennett-Jones, Owen. 2019. "One Screw Short," *London Review of Books,* 18 July, p. 28.
11. Ibid.

12. Ibid, p. 27.
13. Abbas, *Pakistan's Nuclear Bomb,* p. 201.
14. Ibid, p. 8.
15. Abbas, *Pakistan's Nuclear Bomb,* p. 1.
16. Musharraf, Pervez. 2006. *In the Line of Fire. A Memoir*. London: Simon & Schuster, p. 294.
17. Corera, *Shopping for Bombs,* pp. 208-9.
18. Qaiser, Rana. 2004. "Musharraf Pardons Quadeer on Cabinet Nod," *Daily Times,* 6 Feb.
19. Abbas, *Pakistan's Nuclear Bomb,* p. 203.
20. Albright, David and Hinderstein, Corey. 2004. "Uncovering the Nuclear Black-market: Working Toward Closing Gaps in the International Nonproliferation Regime," Institute for Science and International Security, 2 July, http://www.isis-online.org/publications/southasia/nuclear_black_market.html visited November 2019.
21. International Institute for Strategic Studies. 2007. *Nuclear Black Markets: Pakistan, A. Q. Khan and the Rise of Proliferation Networks* (an IISS Strategic Dossier) Ed. Fitzpatrick, Mark. p. 93.
22. Sanger, David E. 2002. "Threats and Alliances; in North Korea and Pakistan, Deep Roots of Nuclear Barter," *The New York Times,* 24 November, Section 1, Page 1.
23. Hersh, Seymour. 2004. "The Deal: Why is Washington Going Easy on Pakistan Nuclear Black Marketers?" *New Yorker,* 8 March, http://www.newyorker.com/fact/content/?040308fa_fact
24. Abbas, *Pakistan's Nuclear Bomb,* p. 7.
25. Pardon for scientist who sold nuclear secrets. February 6, 2004, *The Irish Times.*
26. Albright, David and Hinderstein, Corey. 2005. "Unraveling the A. Q. Khan and future proliferation networks," *The Washington Quarterly,* Vol. 28, N. 2, Spring, p. 111.
27. Potter William J., Sanger, David, and Bunner, Raymond. 2004. "How Pakistani's Network Offered the Whole Kit," *New York Times,* 13 February; Albright, David. 1995. "An Iranian Bomb," *Bulletin of the Atomic Scientists,* V. 51, N. 4, July/August 1995, pp. 20-26.
28. Albright and Hinderstein, p. 121.
29. Bureau of Nonproliferation, US Department of State. 2004. "The Proliferation Security Initiative," July 28, https://www.state.gov/about-the-proliferation-security-initiative
30. Shaikh, Farzana. 2002. "Pakistan's nuclear bomb: beyond the non-proliferation regime," *International Affairs* (London), January 2002, Vol. 78 (1), pp. 29-48.
31. Gregory, Shaun. 2019. "The Terrorist Threat to Pakistan's Nuclear Weapons," CTS Sentinal, Vol. 2, Issue 7, The Combating Terrorism Center.cts.usma.edu/the-terroris-threat-to-pakistans-nuclear-weapons, visited Nov. 2019.
32. Shaikh, p. 48.

TWELVE

The Ultimate Weapon as the Ultimate Guarantee of Regime Survival

North Korea and the Bomb

Proliferation of weapons of mass destruction and advanced delivery systems has emerged as the most important threat to security on the Korean peninsula as well as to international security.[1] The Democratic People's Republic of Korea (DPRK) is a largely isolated state, and a culture preoccupied with eventually ruling Korea as a single nation, within a militaristic posture and with absolutist control.[2] North Korea's drive to defend itself militarily may have been a symptom of their feelings of insecurity and was aimed at the fundamental goal of regime preservation. The North's enduringly hostile relations with the United States and the Republic of Korea (ROK), and its unwillingness simply to rely on either China or the Soviet Union for its security made the ultimate weapon into the ultimate guarantee of regime survival.[3] The North wanted a nuclear arsenal to deter the United States' potential use of nuclear weapons in the region and to protect itself as an outlier from the US nuclear umbrella. The United States and South Korea have a long track record of planning and expressing a willingness to pursue regime change or forced unification of the peninsula, which would effectively erase the state of North Korea as we know it.[4]

Due in part to Moscow's establishment of diplomatic relations with South Korea the North began to doubt the reliability of its traditional military ally and source of nuclear protection.[5] A nuclear North Korea has the ability to make sure that the price of eliminating the regime would be the world's first nuclear war between nuclear states.[6]

North Korea may have also viewed a nuclear arsenal as insurance against an eventual South Korean conventional military supremacy.[7] It may have wanted nuclear weapons because they offered them diplomatic leverage, forcing the world to take more seriously their concerns and promoting direct bilateral talks with the United States.[8] Possession of nuclear weapons would allow the North to reduce its dependence on China and Russia and to increase its freedom of independent action.[9]

The first, and potentially most significant, influence on Pyongyang's nuclear thinking, came from US threats to use nuclear weapons against the North in defense of the South. During the Korean War the United States made a series of implied and direct nuclear threats in its efforts to discourage North Korean and Chinese aggression.[10]

Since it was created in 1968, the Nuclear Non-Proliferation Treaty has played an important role in spreading the norm of nuclear nonproliferation and in preventing non-nuclear states from developing nuclear weapons.[11] However, the NPT has also shown notable weaknesses as three states, India, Pakistan, and Israel, have never signed the treaty but have developed nuclear weapons outside the treaty's jurisdiction. In addition, North Korea, which did sign the treaty in 1985, withdrew from it in 2003 and has since become a de-facto nuclear power. This may have set a precedent for other countries in similar situations.[12]

For a quarter of a century the DPRK has defied, stymied, deferred or circumvented repeated efforts by its allies, its adversaries and the International Atomic Energy Agency (IAEA) to inhibit its pursuit of nuclear weapons and ensure Pyongyang's compliance with its declared non-proliferation obligations. The DPRK signaled its final breach of the nuclear divide with its pull-out from the Nuclear Non-Proliferation Treaty (NPT).[13] Since its withdrawal from the NPT in January 2003, North-Korea has enjoyed the freedom to conduct underground nuclear tests.[14] At the present time, in theory the United States could locate all of North Korea's nuclear-armed missiles and launch a disabling first strike that would wipe them out before they could be launched. But by diversifying the systems it uses to deliver their nuclear weapons, which include traditional missile sites, but also mobile launchers, and roving submarines, North Korea has made its nuclear weapons more difficult to detect and to destroy. North Korea has also begun using solid fuel, as opposed to only liquid fuel to prepare missiles that they would launch making them much harder to detect beforehand and reducing the amount of time needed for repeated launches. Thus, a first-wave US attack to eliminate North Korea's nuclear arsenal gradually becomes impossible as the North's nuclear and missile inventory grow. The North now depends more on mobile missile launchers and disperses its nuclear and missile forces into a larger area.

This means that a survivable nuclear arsenal would guarantee that North Korea could turn any war, invasion, or attempt at regime change into a nuclear conflict. If the regime were to believe that its days were numbered, there would be no incentive for them to refrain from going nuclear. Thanks to the way North Korea thinks about the political value of using force, it has not only the ability, but also the motivation to resort to the use of nuclear weapons. The only uncertainty is the set of concrete circumstances that would lead it to do so. The nuclear weapons in the North create the unprecedented risk that North Korea would engage in nuclear first use, even though the US-South Korea alliance did not intend to pursue regime change or invade Pyongyang.[15] Nuclear weapons make North Korea more capable of pursuing what it has always sought through its foreign policy: improved international standing and the diminution of the US alliance with South Korea so that it might unify the Peninsula on terms favorable to itself.[16]

North Korea knows all it needs to know about the US military plans. North Korean hackers stole alliance war plans in 2016, making the North Korean leader Kim Jong Un fully aware of the alliance's preparations to do what it openly signals it intends to do: dismantle the Kim regime in the event of war.[17]

The crux of the North's threat is its leaders' desire over seventy years to rule the entire Korean peninsula. From the start North Korea's leaders felt that they were required to be threatening and bellicose to survive.[18]

Three questions come to mind in view of the present situation: 1) How will a nuclear armed DPRK likely behave in the future? 2) How will the United States, China and the ROK respond? 3) What will all this mean for peace and stability in North East Asia? The emphasis is increasingly on the military threat of the DPRK, a threat initially conventional and now increasingly becoming nuclear.[19]

The two Koreas remain among the most heavily armed states for major war making in the world, continuously upgrading their military capabilities. In 2016, the North Korean armed forces numbered well over one million, while the ROK's was over six hundred thousand, but their soldiers are much better armed with conventional weapons.[20] In the past two decades, the North has become steadily more threatening and dangerous by developing nuclear weapons and systems for their use at ever-greater distances and because of its extensive participation in proliferation efforts, its dangerous military provocations over the years, its continuing human rights violations, and its flow of refugees.

The NPT system has several fundamental weaknesses. Depending on the intent of users, peaceful nuclear activities can easily be diverted for military purposes. One example is the spread of centrifuge enrichment technology. This technology enables states to produce low-enriched uranium to fuel reactors, but states can use the same technology to make a nuclear bomb by acquiring weapons-grade uranium.[21]

Something like this may have taken place between nuclear scientist Abdul Qadeer Khan and Pyongyang which represented the pivotal turning point in the North's pursuit of nuclear weapons. In the mid to late 1990s, Khan transferred a few dozen gas centrifuges to North Korea. He also facilitated Pyonyang's acquisition of materials, equipment and data design for nuclear enrichment; and granted North Korean personnel access to his own laboratories.

In his official interrogation by Pakistani investigators, Khan asserted that North Korea first built a pilot enrichment plant in the early 1990s, several years before Pakistan purportedly undertook major transactions with the DPRK. He also claimed that North Korean scientists showed him six cores for three plutonium weapons in 1999, seven years before the DPRK detonated its first nuclear device.[22]

Very disturbing for the international community is the continuation of this type of proliferation by Pyongyang and therefore the failure of deterrence with regard to nuclear weapons proliferation by North Korea through its dealings with countries such as Iran, Pakistan, China, and Russia.[23] This transferring of material and technology with nuclear weapons potential to other nuclear aspirants has provoked sharp responses from the international community.[24]

However, the most consequential problem has been its nuclear weapons effort. Estimates of the number of North Korea's nuclear warheads now range as high as twenty to thirty.[25] As of November 30, 2017, North Korea had carried out 117 tests of strategic missiles since its first test in 1984. Under Kim Jong-un, more than eighty strategic missile tests have been undertaken.[26] The US government estimated in 2017 that the country may be producing enough nuclear material each year for twelve additional nuclear warheads.[27] The North is also attempting to miniaturize these weapons so its latest missiles with a potential range of up to nine thousand kilometers, would be able to strike parts of the United States.[28]

All of these developments constitute reasons for serious concern by the world community. The size of the nuclear detonation on September 3, 2017, which the North claimed was produced by a hydrogen bomb has raised fears of a possible escape of radiation from the underground test facility near Kilju at Punggye-ri in the area of Mantapsan.[29] The fears of the possible escape of radiation were compounded by a series of small earthquakes reported from Kilju that have been attributed to underground cave-ins caused by the powerful test. North Korea's test site at Punggye-ri is surrounded by 2.5 million people and many of them do not even know that the test site existed.

Another notable concern is the bombs' potential to contaminate the area with radioactive material. North Korea's mountain test site has collapsed, putting China and other nearby nations at unprecedented risk of radioactive exposure, two separate groups of Chinese scientists studying the issue have confirmed.

The collapse after five nuclear blasts may be why North Korean leader Kim Jong-un declared on Friday, November 23, 2017, that he would freeze the North's nuclear and missile tests and shut down the site, one researcher said. As of the beginning of 2020, this freeze has ended. The last five of Pyongyang's six nuclear tests have all been carried out in North Korea's northwest.

One group of Chinese researchers found that the most recent blast tore open a hole in the mountain, which then collapsed upon itself. A second group concluded that the breakdown created a "chimney" that could allow radioactive fallout from the blast zone below to rise into the air.

A research team led by Wen Lianxing, a geologist with the University of Science and Technology of China in Hefei, concluded that the collapse occurred following the detonation in the fall of 2017 of North Korea's most powerful thermal nuclear warhead in a tunnel about 700 meters (2,296 feet) below the mountain's peak. The test turned the mountain into fragile fragments, the researchers found. There are fears, therefore, that Pyongyang's latest nuclear test may have caused a fallout leak.[30]

Radiation would also have an impact on other forms of life in addition to humans. "In areas where humans are killed or injured by radiation, the same lethality for animals would be expected."[31] The authors of the book *Effects of Nuclear Earth-Penetrator and other weapons* noted that plants would also get hit hard, especially pine and spruce, which are among the most sensitive to radiation.[32] The area around the test site in North Korea is, in fact, becoming a wasteland according to four persons who have defected from the region. They reported that many of the trees planted in the area die, and underground wells have run dry. A northern defector who used to work as a forestry official in Kilju said, "If you plant trees in the mountains, 80 percent of them die. You can blame it on poor planting but the number of trees that die is higher than in other mountains." The Research Association of Vision of North Korea, a defector-related group, has interviewed twenty-one defectors who used to live in Kilju in the last few years. These North Korean defectors claim that in addition to the trees that are planted which die, babies are being born with defects. "I heard from a relative in Kilju that deformed babies were born in hospitals there," one defector said.[33] Another said "I spoke on the phone with family members I left behind and they told me that all of the underground wells dried up after the sixth nuclear test."

Suh Kyun-ryu, a professor of nuclear engineering at Seoul National University, said "Due to the collapsed ground layer, fissures must have formed underneath leading to contamination of the underground layer and water supply."[34] Other defectors said local delicacies such as trout and pine mushrooms have disappeared from the region.

Officials are trying hard to prevent accounts from inside Kilju spreading to other parts of the North. One defector said, "People who boarded trains to the border camps with samples of soil, water, and leaves from Kilju county were arrested and sent to prison camps."

In South Korea, the Unification Ministry has been conducting radioactive contamination tests of thirty North Korean defectors since November 2017.[35]

Unfortunately, the time to prevent the DPRK from developing nuclear weapons capable of hitting the United States and its East Asian allies is virtually over; policy makers and scholars must now turn their attention to the question of how to deter a nuclear North Korea since on many levels the goal of non-proliferation on the Korean peninsula has failed. In 2017, for the first time since the Cold War, a US president found himself explicitly, publicly, and repeatedly threatening war against a nuclear armed adversary. On August 8, 2017, the *Washington Post* published part of a leaked Defense Intelligence Agency assessment with an alarming conclusion: North Korea could now arm ballistic missiles capable of reaching the United States with nuclear warheads.[36] In fact, the DPRK will soon be in a position to deter the United States, Japan and the ROK with its nuclear arsenal, nonproliferation and disarmament increasingly appear to be unrealizable goals now on the Korean peninsula.[37]

NOTES

1. Graybeal, Sidney, and McFate, Patricia. 1992. cited in *Shaping Nuclear Policy for the 1990s; a compendium of Views.* Washington, DC: Report of the Defense Policy Panel, House Armed Services Committee, December 17, p. 280.
2. Kim, Sung Chull. 2017. *North Korea and nuclear weapons: entering the new era of deterrence.* Washington DC: Georgetown University Press, p. 16.
3. Jackson, Van. 2019. *On the Brink Trump, Kim, and the Threat of Nuclear War.* Cambridge: University Press, p. 23.
4. Jackson, Van, p. 42.
5. Ibid, p. 17.
6. Jackson, Van, Ibid.
7. Mach, Andrew. 1991. "North Korea and the Bomb," *Foreign Policy,* No. 83 (Summer 1991), 93-94.
8. *Shaping Nuclear Policy,* p.18.
9. Spector, Leonard and Smith, Jacqueline. 1991., "North Korea: The Next Nuclear Nightmare?" *Arms Control Today,* Vol. 21, Number 2 (March 1991), 10.
10. *Shaping Nuclear Policy,* p. 20.
11. Hilborne, Mark 2012. "The NPT" in *Handbook of Nuclear Proliferation,* ed. Pant, Hersh V. New York: Routledge, 2012, p. 255.
12. Ibid, p. 179.
13. Pollack, Jonathan D. 2011. *North Korea, Nuclear Weapons and International Security,* Oxon: The International Institute for Strategic Studies, p. 13.
14. Ibid.
15. Jackson, Van, p. 43.
16. Ibid, p. 50
17. Shieber, Jonathan. 2017. "North Korean Hackers Stole South Korean and US War Plans," *TechCrunch,* October 10, https://techcrunch.com/2017/10/10/report-north-korean-hackers-stole-south-korean-and-u-s-war-plans visited November 2019.
18. Kim, Sing Chull, 2017. *North Korea and nuclear weapons: entering the new era of deterrence.* Washington, DC: Georgetown University Press; Morgan, Patrick. "Nonproliferation or Deterrence or Both," p. 16.

19. Ibid.
20. International Institute for Strategic Studies (IISS), 2016. *The Military Balance 2016*, London: Routledge, pp. 264-70.
21. Gilinsky, Victor, and Sokoloski, Henry. 2014., "Serious Rules for Nuclear Power without Proliferation," *Nonproliferation Review*, 21 (2014): 79.
22. Pollack, p. 135.
23. Kim, Sung Chull, pp. 20-21.
24. Pollack, p. 14.
25. Arms Control Today, armscontrol.org/factsheets/northkoreaprofile visited January 2020.
26. Sung-hun, Choe, Sanger, David . 2017. "North Korea Fires Missile over Japan," *The New York Times*, 28 Aug., www.nytimes.com/2017/08/28/world/asia/north-korea-missile.html
27. Pand, Ankite. 2017. "North Korea May Already be Annually Accruing Enough Fissile Material for 12 nuclear weapons," *The Diplomat*, August 2017. www.thediplomat.com visited January 2020
28. IISS, Strategic Survey 2015, Oxon: Routledge, 2015, pp. 367-68.
29. Sung-Han, Choe. 2017. North Korean Defectors Show Signs of Possible Radiation Exposure, 27 Dec. Nytimes.com/2017/12/27/world/asia/north-korea-nuclear-test-radiation.html, visited January 2020.
30. Chen, Stephen. 2018."North Korea's nuclear test site has collapsed . . . and that may be why Kim Jong-un suspended tests," *South China Morning Post"* April 25, scmp.com/news/china/diplomacy-defence/article/2143171/north-koreas-nuclear-test-site-has-collapsed-and-may-be-why-kim-jong-un-suspended-tests, visited Jan 2020.
31. Effects of Nuclear Earth-Penetrator and other weapons, 2005. Committee on the Effects of Nuclear Earth-Penetrator and Other Weapons, Division on Engineering and Physical Sciences, The National Academies Press, Washington DC, nap.edu/read/11282/chapter 1, p. 92.
32. Ibid, p. 93.
33. Myon-son, Kim, Libo, Chosun. 2017. "N. Korean Test Site 'Heavily Contaminated'" (English edition), chosun.com/site/data/html_dir/2017/11/06/2017110601163.html.
34. Ibid.
35. Sang-Hun, Choe. 2017. "North Korean Show Signs of Possible Radiation Exposure," *New York Times*, Dec. 27, nytimes.com/2017/12/27/world/asia/north-korea-nuclear-test-radiation.html visited January 2020.
36. Jackson, Van, p. 1.
37. Ibid, p. 3.

Conclusion

Some claim that nuclear weapons keep us safe, but, in fact, the opposite is true. Along with the ever-present threat of nuclear war, there have been many accidents or near misses due to human, technical and computer errors. The Red Cross acknowledges that responding meaningfully to a nuclear detonation is next to impossible.[1]

The year 2019 saw the most world-wide spending on nuclear weapons in history.[2] The nine nuclear-armed countries spent $72.9 billion on their 13,000-plus nuclear weapons in 2019, a $7.1 billion increase from 2018.

One large thermonuclear blast would produce a massive firestorm leaving an area tens of kilometers wide starved of oxygen and killing any survivors of the initial blast. Even a limited regional nuclear war, for example between Pakistan and India, would produce enough soot to trigger a decade long nuclear winter. Reduced crop yields would translate into global famine.[3]

The year 2020 marked the fiftieth anniversary of the Nuclear Non-Proliferation Treaty (NPT). With 191 states party to the NPT, this treaty is the cornerstone of the global regime for nuclear non-proliferation and peaceful uses of nuclear energy.[4] On balance, the NPT has been successful.

However, there are still 13,890 nuclear weapons, 3,750 actively deployed and 1,800 on accident-prone high alert, ready to be launched within minutes.[5] Current arsenals are big but they are only about one-fifth the size of what they were a half century ago. Today there are nine countries with nuclear weapons, the same number as twenty-five years ago, which is a noteworthy indication of the NPT's durability as well as its contribution to international stability. While super power arms control, and the end of the Cold War, deserve most of the credit, along with restraint by China, still the NPT created some of the broader political context and moral pressure that led to these reductions. While it sought to prevent non-nuclear states from ever getting the bomb, its main point is the bargain requiring the existing nuclear weapons states to reduce and ultimately eliminate their arsenals. As a result, of various arms reduction agreements and unilateral decisions, both Washington and Moscow have dramatically cut back on their nuclear arsenals from peaks of over 30,000 and 40,000 respectively to active arsenals of under 4,500 weapons each. The nuclear stores of the United States and Russia, however, still make up 90 percent of the total number of warheads on Earth today. The United Kingdom, France, China, India, and Pakistan each likely possessing between 150 and 300, with Israel's unconfirmed arsenal totaling almost 100 bombs. North Korea probably has several dozen nuclear weapons with enough fissile material to make a few dozen more.[6]

The lack of recent progress by the nuclear weapons states to reduce further their arsenals—with the NPT's ultimate goal of complete nuclear disarmament not on the horizon-frustrates many non-nuclear weapons states. This has given birth to the 2017 Treaty on the Prohibition of Nuclear Weapons (TPNW).[7]

The TPNW explicitly prohibits under all circumstances the development, production, stockpiling, transfer, and use of nuclear weapons, and other nuclear devices. Hosting and threatening to use nuclear explosive devices are also illegal. The TPNW reaffirms that the full and effective implementation of the Treaty on the Non-Proliferation of Nuclear Weapons has a vital role to play in promoting international peace and security. Another vital treaty, which is a core element of nuclear disarmament and the non-proliferation regime, is the Comprehensive Nuclear Test-Ban Treaty and its verification regime.[8]

Unfortunately, the dismantling of other agreements that also slowed the arms race has occurred. Without these agreements, Russia and the United States might now have 65,000 warheads instead of 13,000 at present. Beginning with President George W. Bush's withdrawal from the Anti-Ballistic Missile Treaty in 2001, this dismantlement continued in more recent years by the fact that Donald Trump has wiped away almost all the others that were still in place. In 2018, Trump announced that the United States would withdraw from the Intermediate Nuclear Forces (INF) Treaty. The Administration then announced its intention of leaving the 1992 multi-lateral Open Skies Treaty, which allows those who sign it to fly unarmed observation flights over the territory of others to study military forces and activities. Though satellite technology has lessened its value, it is still important to European parties and has been important to strategic stability.[9]

The only remaining limit on strategic arms is New START, which is due to expire two weeks after the next President of the United States is inaugurated, unless extended by mutual agreement for five more years. The treaty limits each side to 1,550 deployed strategic warheads and 700 launchers. The United States now insists that it will not extend the treaty unless China is included.[10] Beijing will not likely become involved.

The TPNW points out that atomic radiation is more harmful to women than to men.[11] A woman is at a significantly greater risk of suffering and dying from radiation–induced cancer than a man who gets the same dose of Ionizing Radiation. The National Academy of Science report finds that radiation kills men, but it kills significantly more women. The harm to women, such as cancer, is 50 percent higher than the comparable harm to men from radiation doses that fall within the legal limit to the public over a lifetime.[12]

Studies by some of the world's best climate scientists in 2007 found that even a small-scale regional nuclear war would have severe global consequences. One hundred Hiroshima- sized bombs targeted on cities in a war between India and Pakistan, constituting less than half those nations' current nuclear arsenals, would send up 6 million tons of smoke into the stratosphere and beyond.[13] One hundred Hiroshima-sized bombs represents less than 0.5 percent of the global nuclear arsenal and less than 0.1 percent of its explosive yield. For this reason each of the smaller nuclear arsenals of China, France, India, Israel, North Korea, Pakistan and the UK constitutes a global threat, and not only the massive arsenals of Russia and the United States even though they represent most of the world's nuclear weapons.[14]

Six million tons of high-altitude smoke would block sunlight, while at the same time rapidly cooling, darkening and drying the climate worldwide. While temperatures on the Earth's surface would fall on average by 1.5 degrees Celsius, the smoke would heat the upper atmosphere by 50 degrees to 80 degrees Celsius, dramatically reducing the stratospheric ozone which protects us from biologically harmful ultraviolet (UV) radiation. UV levels around the globe would grow to levels not seen before. This would negatively affect our eyes, skin, and immune systems as well as badly damaging plants and animals on land and in the water, including the plankton, which is the basis of the marine food chain.[15]

In the continental interiors where the majority of the global grain is grown, temperatures would go down by 5 degrees to 8 degrees Celsius, with more frequent frosts and growing seasons averaging one month or 25 percent shorter. These colder, darker and drier conditions would reduce major grain crops in the United States and China, the world's largest producers, by 15 percent to 40 percent in the first five years and 10–25 percent for the second five years.[16] This would affect everywhere. At higher latitudes, food production would basically cease.

These are conservative estimates because they do not yet include the effects of increased ultraviolet radiation, radioactive and chemical contamination of productive land or the disruption to the complex global distribution system for water, fertiliser, seed pesticides, and fuel.

The UN Food and Agriculture Organization estimates that 821 million people are chronically malnourished today.[17] More than 300 million additional people are dependent for more than half their energy intake on imported food. Global grain stocks generally provide for between 60 and 115 days' consumption, and production is becoming more erratic as the effects of climate disruption accelerate. Humanity could not withstand sustained declines in food production of this magnitude and as many as 2 billion people would face death from starvation from a "small" regional war anywhere in the world.[18]

Let us leave the concluding words on this topic to the World Health Organization, which has said:[19]

> Nuclear weapons constitute the greatest immediate threat to the health and welfare of humankind, and that no health services in the world could meaningfully respond to the vast number of casualties from even a single nuclear explosion over a city.

NOTES

1. Eide, Espen Barth. 2013. 2010 Review Conference of Nuclear Non-Proliferation Treaty, Chair's summary, Humanitarian impact of nuclear weapons, Oslo, Norway, March 4-5, www://regjeromgem.no/en/archive/Stoltenbergs-2nd-Government/Ministry-of-foreign-affairs-taler-ogartikler-2013/nuclear_summary.hthl?id=716343
2. "Enough is Enough: 2019 Global Nuclear Weapons Spending," International Campaign to Abolish Nuclear Weapons (Geneva) Op. Cit..
3. Barrie, John, 2014. "Improving UN planning for a humanitarian response to a nuclear detonation," *Bulletin of the Atomic Scientists*. 70(5), 74-85.
4. Rauf, Tariq. 2020. "The 2020 NPT Review Conference From the Sublime to the Ridiculous," *Inter Press News Agency*, Friday, March 20..
5. Kristensen, Hans M. 2019. "Status of world nuclear forces," *Federation of American Scientists*, May, 2019, https://fas.org/issues/nuclear-weapons/status-world-nuclear-forces, p. 29.
6. O'Hanlon, Michael E., Einhorn, Pifer, Dteven, and Ross, Frank A. 2020. "Order from Chaos. Experts assess the nuclear Non-Proliferation Treaty, 50 years after it went into effect." *Brookings*, March 3rd.
7. Choosing Humanity Why Australia Must Join the Treaty on the Prohibition of Nuclear Weapons. Icanw.org.au/up-content/uploads/Choosing-Humanity-ICAN-Report.pdf
8. Treaty on the Prohibition of Nuclear Weapons. Treaties.un.org/doc/Treaties/2017/07/20170707?2003-42?20PM/CH_XXVI_9.pdf
9. Matthews, Jessica. 2020. "The New Nuclear Threat," *New York Review of Books* p. 21.
10. Ibid.
11. World Information Service on Energy. 2011. Nuclear Monitor Issue #736 Number 6192, 11/11/2011 NIRS Briefing Paper
12. World Information Service on Energy. 2011. Nuclear Monitor Issue #736, Number 6192, 11/11/2011, NIRS, Briefing Paper.BEIRVII 2006., Phase 2 report, "Health Risks from Exposure to Low Levels of Ionizing Radiation," Natopma; Academy pf Sciences.
13. Toon, Owen, Bardeen, Charles, Robock, Alan, Paterson, R.J., Xia, Lili. 2018. "Rapid expansion of nuclear arsenals by Pakistan and India threatens regional and global catastrophe," *American Geophysical Union*, Fall Meeting, Washington, DC, Dec. 2018, GC33B-12.
14. Holfand, Ira, Haines, Andy, Ruff, Tilman, Kristensen, Hans M., Lewis, Patricia, Mian, Zia. 2018. "The Growing Threat of nuclear war and the role of the health community," *World Medical Journal*, 62(3), 86-94.
15. Mills, Michael, Toon, Owen, Taylor, Julia-Lee, Robock, Alan. 2015. "Multi-decadal global cooling and unprecedented ozone loss following a regional nuclear conflict," *Earth's Future* 2:161-76, pp. 29-34.
16. Toon, Owen, Robock, Alan, Mills, Michael, Xia, Lili. 2017. "Asia treads the nuclear path, unaware that self-assured destruction would result from nuclear war," *J. Asian Studies*, 76:437-56.

17. FAO, IFAD, WFP and WHO. 2018. The State of Food Security in the World. Building climate resilience for food security and nutrition, FAO.

18. Holland, Ira. 2013. *Nuclear Famine. Two Billion People at Risk*. Boston: International Physicians for the Prevention of Nuclear War; Helfand et alii, op. cit., "The Growing Threat of Nuclear War."

19. World Health Organization. 1984 and 1987. *Effects of Nuclear War on Health and Health Services*. Geneva.

Bibliography

Abbas, Hassan, *Pakistan's Nuclear Bomb: A Story of Defiance, Deterrence and Deviance.* Oxford: University Press, 2018

Akleyev, Alexander V. "The Long Shadow of Soviet Plutonium Production," *Environment* Vol. 39, No. 1, June 1997, pp. 12–20

Albright, David, "South Africa and the Affordable Bomb," *Bulletin of the Atomic Scientists,* Vol. 50, No. 4, 1994, pp. 37–47.

———. "Bin Laden and the Bomb," *Bulletin of the Atomic Scientists,* Jan/Feb. 2002, Vol. 58, Issue 1, p. 22.

———. *Peddling Peril, How the Secret Nuclear Arms Trade Arms America's Enemies.* New York: Free Press (Simon & Schuster), 2010, Chapter 8.

———. "Al Qaeda's Nuclear Program: Through the Window of Seized Documents," Nautilis Institute Special Forum, Vol. 47/November 6, 2002.

———. "Uncovering the Nuclear Blackmarket: Working Toward Closing Gaps in the International Nonproliferation Regime," Prepared for the Institute for Materials Management 45th Annual Meeting, Orlando, Florida, July 2, 2004.

———. "Unraveling the A.Q. Khan and future proliferation networks," *The Washington Quarterly,* Vol. 28, No. 2, Spring 2005, p. 111.

Alexievich, Svetlana, *Voices from Chernobyl* London: Dalkey Archive Press, 1997.

Alonso-Perez, S. "Arrival of radionuclides released by the Fukushima Nuclear Power Plant (FDNPP) accident to Tenerife (Canary Islands), *Journal of Environmental Radioactivity* Vol 116, Feb. 2013, pp. 180–86.

Aoyama, Michio, "Distribution of 137 Cs and 134 Cs in the North Pacific Ocean: impacts of TEPCO Fukushima Daichi NPP Accident" *Journal of Radioanalytic and Nuclear Chemistry* Vol. 296, 2013, p. 535.

Appadori, A, *India's Foreign Policy and Relations,* New Delhi: South Asian Private, 1985.

Apsalikov, Kazbek, "Radiation Exposure due to local fallout from Soviet atmospheric weapons testing in Kazakhstan; solid cancer morbidity in the Semipalantinsk historic cohort 1960-1999, *Radiation Research* Oct. 2005, Vol 65, (4 Pt 1) pp. 409–19.

Bader, William B. *The United States and the Spread of Nuclear Weapons,* New York: Pegassus: 1968.

Ball, S.L. "Military Relations between the United States and Great Britain under the terms of the McMahon Act 1946-1958," *The Historical Journal,* Vol. 38, 2, June 1995.

Bardeen, Charles, "Rapid expansion of nuclear arsenals of Pakistan and India threatens regional and global catastrophe," *American Geophysical Union* Fall Meeting, Washington DC, Dec. 2008.

Barker, Holly M., *The Consequential Damage of Nuclear War, The Rongelap Report.* Walnut Creek: Left Coast Press, 2008.

Barrie, John, "Improving UN planning for a humanitarian response to a nuclear detonation," *Bulletin of the Atomic Scientists* 70(5), 2014.

Batcher, Robert Simon, "The Consequences of an Indo-Pakistani War," *International Studies Review,* Vol. 6, Number 4, December 2004.

Bauer, Suzanne, "Radiation exposure due to local fallout from Soviet atmospheric nuclear weapons testing in Kazakstan, solid cancer morbidity in the Semipalantinsk historial cohort, 1960 1999," *Radiation Research* October 2005, Vol. 164.

Baum, Seth, "Why catastrophes can change the course of history," Global Catastrophic Risk Institute www.bbc.com/future/story/20190408-how-catastrophes-can-change-the path-of-lear humanity 2019."

Bellovin, Steven, "Permissive Action Links, Nuclear Weapons and the History of Public Key Cryptography," Department of Computer Science, Columbia University, smb@cs.columbia.edu, http:www.u.columbia.edu 2005.

Bennett-Jones, Owen, "One Screw Short," *London Review of Books,* July 18, 2019.

Bevelacqua, J. J., "The Fukushima Accident and its Radiological Impact on the Environment," *The Physics Teacher* Vol. 50 (6), September 2012.

Billiet, Edward, Oversight Hearing before the Subcommittee on Natural Resources, House of Representatives, 103rd Congress "Radiation Exposure from Pacific Nuclear Tests Oversight Hearings," Serial No. 103–68, Feb. 24, 1994.

bin Laden, Osama, Interview, *Time,* December 23, 1998.

Bluth, Christoph, "Review of *Smuggling Armageddon: The Nuclear Black Market in the Former Soviet Union* by Renssalaer Lee, *Slavic Studies* Vol. 59, No. 2, 2000. http://www.bbc.com/future/story/20190408-how-catastrophes-can-change-the path-of-humanity, 2019.

Bouville, Andre, "Radiation doses to local populations near nuclear test sites worldwide," *Health Physics* 82(5), 2002.

———. "Fallout from Nuclear Weapons Tests and Cancer Risks," *American Scientist* Vol. 94, No. 1, January 2006.

Brodie, Bernard (ed.), *The Absolute Weapon: Atomic Power and World Order.* New York: Harcourt Brace, 1946.

Brown, Kate, *Plutopia: Nuclear Families, Atomic Cities, and the Great Soviet and American Plutonium Disasters.* New York: Oxford, 2013.

———. *Manual for Survival A Chernobyl Guide to the Future.* New York: Norton, 2019.

Buehler, Kathryn, "Bin Laden and the Bomb," *Bulletin of the Atomic Scientists,* Vol. 58, Issue 1, Jan/Feb 2002.

Bukharin, Oleg, "Potatoes Were Guarded Better," *Bulletin of Atomic Scientists,"* Lix, Vol. 59, No. 3, 2000.

Bulgac, Aurel, "Nuclear Fission Dynamics: Past, Present Needs and Future," *Frontiers in Physics* 8:63, March 18, 2020.

Bunner, Raymond, "How Pakistan's Network Offered the Whole Kit," *New York Times,* February 13, 2004.

Burke, Taylor, "Nuclear Energy and Proliferation Problems, Observations and Proposals," Federal Energy and Resource Policy Seminar, Dec. 3, 2004, https://lawbepress.com/cgi/viewcontent.cgi?referer=https:www.google.com/&jttparenedin=1&article=4234&context=expresso.

Bustos, J. J., "Arrival of radionuclides released by the Fukushima accident to Tenerife (Canary Islands)," *Journal of Environmental Radioactivity,* Vol. 116, Feb. 2013.

Caldicott, Helen, *Crisis without End: The Medical and Ecological Consequences of the Fukushima Nuclear Catastrophe* [From the Symposium at the New York Academy of Medicine, March 11-12, 2013], New York: The Free Press, 2014.

Carson, Mark. "Can Terrorists Build Nuclear Weapons?" *Nuclear Control Institute,* Washington DC, 2006 www://nci.org/k-m/makeab.htm.

Carter, Ashton B. *Preventive Defense, A New Security Strategy for America.* Washington: Brookings Institute, 1999.

Chen, Stephen, "North Korea's nuclear test site has collapsed . . . and that may be why Kim Jong-un suspended tests," *South China Morning Post,* April 25, 2018.

Corera, Gordon, *Shopping for Bombs Nuclear Proliferation, Global Insecurity and the Rise and Fall of the A.Q. Khan Network.* New York: Oxford University Press, 2006.

Chouzou, Tetsuji, "The Influence of the Lop Nor Nuclear Weapons Test Base on the population of Kazakhstan," *Radiation Measurements,* Vol. 46(4), 2011.

Claque, David A. (Ed.), *Encyclopedia of Islands,* Berkeley CA: University of California Press, 2009.

Collins, Daniel, "Nuclear Accidents in the former Soviet Union: Kyshtym, Chelyabinsk and Chernobyl," Defense National Institute, Uniformed Service University (USU), DNA/A FRRI 4020, ADA 254 669.

Connor, Tim, "Nuclear Workers at Risk," *Bulletin of the Atomic Scientists*, Vol. 46, No. 7, Sept. 1990.
Cuevas, E., "Arrival of radionuclides released by the Fukushima accident at Tenerife (Canary Islands)," *Journal of Environmental Radioactivity*, Vol. 116, Feb. 2013.
Daly, Sara, Project Air Force (US), Rand Corporation, *Aum Shinrikyo, Al Quaeda, and the Kinshasa Reactor: Implications for combating nuclear terrorism*, 2005.
Danielsson, Beng, "Poisoned Pacific: The Legacy of French Nuclear Testing," *Bulletin of the Atomic Scientists*. Vol. 46, No. 2 (March 1990).
Darby, S. C. "A summary of mortality and incidence of cancer in men from the United Kingdom who participated in the United Kingdom's atmospheric nuclear tests and experimental programs," *British Medical Journal* 296, Issue 6618, 1988.
Dauer, L. T., "The Japanese Tsunami and Resulting Nuclear Emergency at the Fukushima Daichi Power Facility: Technical, Radiologic and Response Perspectives," *Journal of Nuclear Medicine* 52, Sept. 2011.
Davies, Hugh, "Bin Laden Aide Tried to Buy Atomic Arms," *Daily Telegraph* (London) September 1998.
DeGroot, Gerard, *The Bomb. A Life*. Cambridge MA: Harvard University Press, 2006.
Delacroix, D. *Radionuclide and Radiation Protection Handbook*. Nuclear Technology Publishing ISBN 978-1870965873, 2002.
Devgan, Jas, "The Hanford site historic district: History of the plutonium facilities 1943-1990," *Environmental Progress*, Vol. 22 (4), Dec. 2003.
Devine, R. A., "Atoms for Peace and War 1953-1961," Review *Science* 10 Nov. 1989, Vol. 246 (4931).
Dinara, Stephanenko, "The Influence of the Lop Nor Nuclear Weapons Test Ban on the population of the Republic of Kazakhstan," *Radiation Measurements*, Vol. 46 (4), 2011.
Divine, Robert, *Blowing on the Wind. The Nuclear Test Ban Debate 1954-60*. New York: Oxford University, 1978.
Duarte-Rodriguez, X. "Arrival of radionuclides released by the Fukushima accident to Tenerife (Canary Islands)," *Journal of Environmental Radioactivity*, Vol. 116, Feb. 2013.
Dunning, Gordon M., "The Effects of Nuclear Weapons Testing," *Scientific Monthly*, Vol. 31, No. 6 (December 1995).
Dyer, Gwynne, *War*. The New Edition. Toronto: Random House, 2016.
Eide, Espan Barth, "Review Conference of Nuclear Non-Proliferation Treaty, Chair's Summary," Humanitarian impact of nuclear weapons, Oslo, Norway, 4–5 March 2013.
Einhorn, Pilfer Dteven, "Order from Chaos. Experts assess the nuclear Non-Proliferation Treaty, 50 years after it went into effect," *Brookings Institute* March 2020.
Eisenhower, Dwight D., *Mandate for Change*. Garden City, NY: Doubleday, 1963.
Enomato, Masahiro, "Radiological impact of the nuclear power plant accident on freshwater fish: An overview of monitoring results," *Journal of Environmental Radioactivity* Vol. 151, Part 1, Jan. 2, 2016.
Eyle, Rindzevicute, *The Power of Systems How the Policy Sciences Opened Up the Cold War*.2016 Ebook.
Faisal, M., "Pakistan's Strategic Nuclear Policy and Implications for Deterrence Stability," *CISS Insight Quarterly News and Views*, IV (1) 2016.
Farmelo, Graham, *Churchill's Bomb: A Hidden History of the First Nuclear Weapons Program*. London: Faber & Faber, 2014.
Fell, T. P., "A Summary of mortality and incidence of cancer in men from the United Kingdom's atmospheric nuclear tests and experimental programs," *British Medical Journal*, 296 Issue 6618, 1988.
Fitzpatrick, Mark (ed), *Nuclear Black Markets: Pakistan, A.Q. Khan and the Rise of Proliferation Networks* (Am IISS Strategic Dossier), 2007.
Ganguly, Sumit, "India's pathway to Pokhran II," *International Security*, Vol. 23:4, Spring 1999.

Geller, David S., "Nuclear Weapons and the Indo-Pakistani Conflict: Global Implications of a Regional Power Cycle," *Political Science Review*. Vol. 24, No. 1, Jan. 2003.
Gephart. R. E. "A Conversation about Nuclear Waste and Cleanup," *Journal of Environmental Quality*, Vol. 34 (1), Jan. 2005.
Gerber, Michele Stenehjem, *On the Home Front: The Cold War Legacy of the Hanford Nuclear Site*. Lincoln: University of Nebraska Press, 1992.
Gilinsky, Victor, "Service Rules for Nuclear Power without Proliferation," *Nonproliferation Review*. 21, 2014.
Gillespie, Rosemary G. "Nuclear Bomb Testing," In *Encyclopedia of Islands*. Berkeley CA: University of California Press, 2009.
Gillette, Robert, "Radiation Spill at Hanford: The Anatomy of an Accident," *Science*, Vol. 181, No. 419, Aug. 24, 1973.
Gonglarenko, Roman, Deutsche Welle, "Ukraine's forgotten security guarantee. The Budapest Memorandum." May 12, 2014.
Graybeal, Sidney, *Shaping Nuclear Policy for the 1990s; a compendium of views*. Washington, DC: Report of the Defense Policy Panel, House Armed Services Committee, Dec. 17, 1992.
Gregory, Shaun. "The Terrorist Threat to Pakistan's Nuclear Weapons," *CTS Sentinel*, Vol. 2, Issue 7, 2019.
Gunaratna, Rohan, *Inside Al Qaeda: Global Network of Terror*. New York: Columbia University Press, 2002.
Gundersen, Arnold, "What Did They Know and When?" *Crisis without End: The Medical and Ecological Consequences of the Fukushima Nuclear Catastrophe* [From the Symposium at the New York Academy of Medicine, March 11–12, 2013], New York: The Free Press, 2014.
Gupta, Amit, "Introduction," *India's National Security*, London: Lynne Rienner Publishers, 2000.
Gurriarant, O. R. D. "Transfer factors of cesium made in milk and meat determined from measurements made in France in 2011 following the Fukushima accident," *Radioprotection* Vol. 48 (3), April 2003.
Gusev, Boris, "Radiation Exposure due to local fallout from Soviet atmospheric testing in Kazakhstan; solid cancer morbidity in the Semipalantansk historical cohort 1960-1999s," *Radiation Research* Oct. 2005 Vol. 164 (4Pt1).
Hachiya, Michikiko, *Hiroshima Diary* Chapel Hill: University of North Carolina Press, 1955.
Halperin,Morton, *China and the Bomb*, London: Pall Mall Press, 1965.
Hamajima, Yasoumori, "Distribution of 137 Cs and 134 Cs in the North Pacific Ocean: Impacts of the TEPCO Fukushima Daichi accident," *Journal of Radioanalytic and Nuclear Chemistry*, Vol. 296 (1), July 6, 2015.
Hanford, J., "The Hanford site historic district: History of the plutonium facilities, 1943-1990," *Environmental Progress*, Vol. 22 (4), 2005.
Harden, Blayne, *A River Lost. The Life and Death of the Columbia*. New York: Norton, 1996.
Hatanaki, Takashife. "The Communication of Information Such as Evacuation Orders at the Time of a Nuclear Power Station Accident," *Japan Medical Association Journal* 57(5-6), Dec. 2014.
Helfand, Ira, "The Growing Threat of Nuclear War and the Role of the Health Community," *World Medical Journal* 62(3), 2018.
Hernández-Armas, J., "Arrival of radionuclides released by the Fukushima accident to Tenerife (Canary Islands)," *Journal of Environmental Radioactivity* Vol. 116, Feb. 2013.
Hersh, Seymour, "The Deal: Why is Washington Going Easy on Pakistan Nuclear Black Marketeers?" *New Yorker* March 8, 2004.
Hewlett, Richard G. *The New World*. New York: Simon & Schuster, 1988.
Hickman, P. *Radionuclide and Radiation Protection Handbook*. Nuclear Technical Publisher, *Health Physics* 83(1): 136 ISBN 978-1870965873, July 2002.

Higginbotham, Adam, *Midnight in Chernobyl. The Untold Story of the World's Greatest Nuclear Disaster.* New York: Simon & Schuster, 2019.

Higgins, Holly, "Bin Laden and the Bomb," *Bulletin of the Atomic Scientists,* Jan/Feb. 2002, Vol. 58, Issue 1.

Hilborne, Mark, "The NPT," in *Handbook of Nuclear Proliferation.* New York: Routledge, 2012.

Hinderstein, Corey, "Unraveling the A.Q. Khan and future proliferation networks, *The Washington Quarterly,* Vol. 28, No. 2, Spring 2005.

Helfand, Ira, *Nuclear Famine: Two Billion People at Risk.* International Physicians for the Prevention of Nuclear War, 2013.

Hujama, A., "The Fukushima nuclear accident and the pale grass blue butterfly. Evaluating biological effects of long-term, low dose exposures," *BMC Evolutionary Biology* 13: 168, 2013.

Hutchinson, Robert, "The Struggle for Control of Radioactive Sources," *Jane's Intelligence Review,* March 2003.

Iibo, Chosun, N. Korean Test Site 'Heavily Contaminated, 2017, chosun.com/site/data/html_dir/2017/11/06/2017110601163.html

Ishida, Masanobu, "Radioactive contamination in the Tokyo metropolitan area in the early stage of the Fukushima Daichi Nuclear Power Plant (FDNPP) accident and its fluctuation of five years," PLos ONE, Vol. 12 (11) p.e. 0187687 Nov, 14, 2017.

Ishii, Masami, "The Communication of Information such as Evacuation Orders at the Time of a Nuclear Power Station Accident," *Japan Medical Association Journal,* 57 (5-6) Dec. 2014.

Ivannikov, Alexander, "The Influence of the Lop Nor Weapons Test Ban on the population of the Republic of Kazakstan," *Radiation Measurements,* Vol. 46 (4), 2011.

Izuumi, Shigehiko, "Radiological impact of the nuclear power plant accident on freshwater fish : An overview of monitoring results," *Journal of Environmental Radioactivity,* Vol. 151, Part 1, June 2016.

Jackson, Van, *On the Brink Trump, Kim and the Threat of Nuclear War,* Cambridge: University Press, 2019.

James, Geoffrey, "Review of *Growth and Decay: Prypiat and the Chernobyl Exclusion Zones* by McMillan, David, *Border Crossings,* Vol. 15, 2019.

Jehl, Douglas, "CIA Says Pakistan Gave Iran Nuclear Aid," *The New York Times,* November 24, 2004.

Johnston, Barbara Rose, *The Consequential Damage of Nuclear War: The Ronelap Report.* Walnut Creek: Left Coast Press, 2008.

Josephson, Paul R., *Red Atom Russia's Nuclear Power Program from Stalin to Today.* .Pittsburgh: University of Pittsburgh Press, 2000.

Kan, Naoto, "No Nuclear Power is the Best Nuclear Power," In: Helen Caldicott (ed.). *Crisis without End: The Medical and Ecological Consequences of the Fukushima Nuclear Catastrophe,* New York: The New Press, 2014

Kawata. Gojo, "Radiological impact of the nuclear power plant accident on freshwater fish: An overview of monitoring results," *Journal of Environmental Radioactivity,* Vol. 151, Part 1, Jan. 2016.

Kendall, G. M., "A summary of mortality and incidence of cancer in men who participated in the United Kingdom's atmospheric nuclear weapons tests and experimental programs," *British Medical Journal,* 296, Issue 6618, 1988.

Khan, Feroze H., *Eating Grass: The Making of the Pakistani Bomb.* Stanford, CA: Stanford University Press, 2012.

Kim, Sung Chull, *North Korea and Nuclear Weapons; entering a new era of deterrence.* Washington, DC: Georgetown University Press, 2017.

Koide, Hiroaki, "Living in a Contaminated World," In: Helen Caldicott (ed.), *Crisis without End.* New York: The New Press, 2014.

Kordu, Matt, "Indian nuclear forces," *Bulletin of the Atomic Scientists.* Vol. 74, Issue 6, 2018. Special issue: The verification of arms control agreements.

Kossenko, M. M. "Cancer Mortality among Techa River Residents and their Offspring," *Health Physics* 71, 1996.

"The Growing Threat of nuclear war and the role of the health community," *World Medical Journal,* 62(3), 2018.

Kristensen, Hans M., "Indian nuclear forces, 2018," *Bulletin of the Atomic Scientists.* Vol. 74, Issue 6, 2018. Special Issue: The verification of arms control agreements.

"Global nuclear weapons inventories, 1945-2010," *Bulletin of the Atomic Scientists,* 66:4, 2010.

———. "Status of world nuclear forces," *Federation of American Scientists,* May 2019.

Kutazawa, Kay, "Fukushima in Review: A complex disaster, a disastrous response," *Bulletin of the Atomic Scientists,* March 1, 2012, Vol. 68 (2).

Kuzbek, N. "Radiation exposure due to local fallout from Soviet atmospheric testing in Kazakhstan; solid cancer morbidity in the Semipalantinsk historical cohort, 1960-1999," *Radiation Research,* October 2005, Vol. 164 (4 Pt 1).

Kuznetsov, M., "An analysis of the hydrogen explosion in the Fukushima Daichi accident," *International Journal of Hydrogen Energy,* Vol. 40 (25), July 6, 2015.

Lambeth, Benjamin St., "Nuclear Proliferation and Arms Control Policy," *Orbis* Vol. XIV, No. 2, Summer 1970.

Land, Charles E., "Fallout from Nuclear Weapons Tests and Cancer Risks," *American Scientist* V. 94, N. 1, January 2006.

Landler, Mark, "Pakistan Chief Says It appears Scientists Sold Nuclear Data," *New York Times,* Jan. 24, 2004.

Langenkamp, "Nuclear Energy and Proliferation Problems, Observations and Proposals," Federal Resource Policy Seminar, Dec.3, 2004.

Lankarani, K. B., "From Hiroshima to Fukushima," *Iranian Red Crescent Medical Journal,* 13(8), Aug. 2011.

Lavoy, Peter, "Islamabad's Nuclear Posture: Its Premises and Implementation," In: Sokoloski, Henry D.S. (ed.), *Pakistan's Nuclear Future: Worries Beyond War,* Carlisle, PA: Strategic Studies International, 2008.

Lawler, Maj. General Bruce, "The Black Sea: Center of the nuclear black market," *Bulletin of the Atomic Scientists,* Vol, 67-6, Dec. 2011.

Leader, Stefan, "Osama bin Laden and the Terrorist Search for WMD," *Jane's Intelligence Review,* Vol. 11, No. 6, 1999.

Leblanc, F. *Radionuclide and Radiation Protection Handbook,* Nuclear Technology Publishing, ISBN 978-1870 965 873, 2009.

Leblanc, Renaud F. P., "Transfer factors of cesium in milk and meat determined from measurements in France in 2011 following the Fukushima Daichi accident, *Radioprotection* Vol. 48 (3), April 2013.

Lee, Rensselaer W., *Smuggling Armageddon: The Nuclear Black Market in the Former Soviet Union and Europe.* New York: St. Martin's Press, 1998.

Lee-Taylor, Julie, "A Multi-decadal global cooling and unprecedented ozone loss following a regional nuclear conflict," *Earth's Future* 2(4), 2014.

Levy, Adrian, *Deception Pakistan, the United States and the Secret Trade in Nuclear Weapons,* New York: Walker & Co., 2007.

Lewis, Patricia, "The Growing Threat of nuclear war and the role of the health community," *World Medical Journal* 62(3), 2018.

Libby, Willard F. Murray, "Radioactive Debris from Operation Castle: Worldwide Fallout," New York Atomic Energy Commission, OSTI, GOV, 1955.

Libo, Chosun, "N. Korean Test Site 'Heavily Contaminated,'" 2017. English chosun.com/site/data/htmldir/2017/11/062017110601163.html.

Lifton, Robert Jay, *Death in Life. Survivors of Hiroshima.* Toronto: Random House, 1967.

———. *Destroying the World to Save It. Aum Shinrikyo Apocalyptic Violence and the New Global Terrorism.* New York: Henry Holt, 1999.

Lyubchansky, E. R., "Environmental and Medical Effects of nuclear weapons production in the Southern Urals," *Science of the Total Environment,* Vol. 142, No. 1, 1994.

Mach, Andrew Walker, "North Korea and the Bomb," *Foreign Policy,* No. 83, Summer 1991.
Macfarquhar, Emily, "Benazir and the Bomb," *Alice Patterson Foundation,* Nov. 14, 2019.
Mandelbaum, Michael, "Lessons of the Next Nuclear War," *Foreign Affairs* 79)2), March-April 1995.
Masaharu, Tetsuji, Chouzou, Hoshi, "The Influence of the Lop Nor Nuclear Weapons Test Base on the population of the Republic of Kazakhstan," *Radiation Meaurements* Vol. 46(4), 2011.
McFate, Patricia, *Shaping Nuclear Policy for the 1990s; a compendium of views,* Washington DC: Report of the Defence Policy Panel, House Armed Services Committee, Dec. 17, 1992.
McGrail, Chris, "Missing keys, holes in the fence, and a single padlock: welcome to Congo's nuclear plant," *Guardian,* November 23, 2006.
Medyedev, Zhores, *Nuclear Disaster in the Urals,* New York: Vintage, 1980.
Mian, Zia, "Fissile Material in South Asia: The Implications of the US-India Nuclear Deal," Research report of the International Panel on Fissile Materials (IPFM), September 2006.
Miklian, Jason, "Hot Rod, Poking around for uranium in the world's least secure nuclear reactor," *Financial Post,* June 28, 2013.
Miller, Nicholas I. *Stopping the Bomb. The Sources and Effectiveness of US Nonproliferation Policy.* Ithaca: Cornell University Press, 2018.
Mills, Michael J., "Multidecadal global cooling and unprecedented ozone loss following a regional nuclear conflict," *Earth's Futures100,* Vol. 2, Issue 4, April 2014.
———. "Asia treads the nuclear path, unaware that self-assured destruction would result from nuclear war," *Journal of Asian Studies,* 76, 2017.
Mir, Hamid, "Osama Claims He Has Nukes: If US Uses N-arms It Will Get Same Response," *Dawn,* Nov. 10, 2001.
Mirchandani, G. G., *India's Nuclear Dilemma,* New Delhi: Humanities Press, 1968.
Morgan, Patrick, "Nonproliferation or Deterrence or Both," In: Kim, Sing Chuli (ed.), *North Korea and Nuclear weapons: Entering the New Era of Deterrence,* Washington DC: Georgetown University Press, 2017.
Morishita, Daigo, "Radiological impact of the nuclear power plant accident on freshwater fish. An overview of monitoring results," *Journal of Environmental Radioactivity.* Vol. 151, Pt. 1, Jan. 2016.
Mousseau, Timothy A., "Reply to Andor Pape Møller regarding 'abundance of birds in Fukushima as judged from Chernobyl," by Møller et al. 2012," *Environmental Pollution.* Vol. 169, Oct. 2014.
"The Biological Consequences of Chernobyl and Fukushima," in: *Crisis without End,* ed. Caldicott, H. New York: The New Press, 2014.
"Genetic and Ecological Studies in Animals in Chernobyl and Fukushima," *Journal of Heredity,* Vol. 105, Issue 5, Sept.–Oct. 2014.
Murray, Willard F. "Radioctive Debris from Operation Castle: Worldwide Fallout," New York: United States Atomic Energy Commission, OSTI GOV, 1955.
Musharraf, Pervez, *In the Line of Fire. A Memoir.* London: Simon and Schuster, 2006,"
Myon-Son, Kim, "N. Korean Test Site 'Heavily Contaminated,'" English chosun.com/site/data/html_dir/2017/11/06/2017 110601163.html
Naiyra, R. "Arrival of radionuclides released by the Fukushima accident to Tenerife (Canary Islands)," *Journal of Environmental Radioactivity,* Vol 116, Feb. 2013.
Nesterenko, Alexey, *Chernobyl Consequences of the Catastrophe for People and Environment. Annals of the New York Academy of Science,* Vol. 1181, Boston: Blackwell, 2009.
Nesterenko, Vassily B., *Chernobyl Consequences of the Catastrophe for People and Environment. Annals of the New York Academy of Science,* Vol. 1181, Boston: Blackwell, 2009.
Nohara, C., "The Fukushima nuclear accident and the pale grass blue butterfly: evaluating biological effects of long term low dose exposures," *BMC Evolutionary Biology,* 13: 168, 2013.

Norris. Robert S., "The Soviet Archipelago," *Arms Control Today. Arms Control Association,* 22(1) Jan./Feb. 1992, Loose Nukes Special Issue.

Novikov, Vladimir, "The Long Shadow of Soviet Plutonium Production," *Environment,* Vol. 39, No.1, Jan. 1, 1997.

———. "Present/Science and Technology/Cold War, Science and Space, Nuclear Plant Disasters. 2011, *World History Encyclopedia,* in ERA 9: Promises and Paradoxes, 1945.

Obama, Barack, Speech in Prague, Spring 2009.

O'Hanlon, Michael E., "Order from Chaos. Experts assess the nuclear Non-Proliferation Treaty, 50 years after it went into effect," *The Brookings Institute,* March 3, 2020.

Ojino, Mayo, "The Communication of Information Such as Evacuation Orders at the time of a Nuclear Power Station Accident," *Japan Medical Association Journal,* 57(5-6), Dec. 2014.

Okamoto, Tanaka, "The influence of the Lop Nor Nuclear Weapons Test Base on the population of the Republic of Kazakhstan," *Radiation Measurement,* Vol. 46(4), 2011.

Oren Clary, Christopher, "The A.Q. Khan Network: Causes and Implications," Master's Thesis, Naval Postgraduate School, Monterey, December 2005.

Orjollet, D., "Transfer factors of cesium in milk and meat determined by measurements made in France in 2013 following the Fukushima accident," *Radioprotection,* Vol. 48(3), April 2013.

Otaki, J. M., "The Fukushima nuclear accident and the pale grass blue butterfly: evolving biological effects of long term, low dose exposures," *BMC Evolutionary Biology,* 13: 168, 2013.

Paban, Patit, "Organized Crime from Trafficking to Terrorism," ABC-CLIO, LLC, 2007.

Pais, Abraham, "Niels Bohr's Times," In: *Physics, Philosophy and Policy,* Oxford: Oxford University Press, 2007.

Pan, Esther, "Nonproliferation—The Pakistan Network," *Council on Foreign Relations* Vol. 83, No. 1, Feb. 2004.

Pand, Ankite, "North Korea May Already be Annually Accruing Enough Fissile Material for 12 nuclear weapons," *The Diplomat,* August 2017.

Parache, V. "Transfer factors of cesium in milk and meat determined from measurements made in France in 2011 following the Fukushima Daichi accident," *Radioprotection,* Vol. 48(3), April 2013.

Parachini, John, *Project Air Force (US):* Rand Corporation, *Aum Shinrikyou, Al Qaeda and the Kinshasa Reactor. Implications of three cases for combating nuclear terrorism,* 2005.

Paterson, Alan, "Rapid expansion of nuclear arsenals by Pakistan and India threatens regional and global catastrophe," *American Geophysical Review,* Fall Meeting, Washington DC, Dec. 2018, GC 33B-12.

Perez-Lopez, "Arrival of radionuclides released by the Fukushima accident to Tenerife (Canary Islands)," *Journal of Environmental Radioactivity,* Vol. 15, Feb. 2013.

Perkovich, George, *India's Nuclear Bomb. The Impact on Global Proliferation.* Berkeley: University of California Press, 1999.

Perry, William J., *Preventive Defense. A New Security Strategy for America.* Washington: Brookings Institution Press, 1999.

Pifer, Steven, "Order from Chaos. Why Care about Ukraine and the Budapest Memorandum," *Brookings,* Dec. 5, 2019.

Pinkham, Sophie, "The Chernobyl Syndrome," *New York Review of Books,* April 4, 2019.

Pivinia, Ludmilla M., "Radiation exposure due to local fallout from Soviet atmospheric weapons testing in Kazakhstan; solid cancer morbidity in the Semiplantinsk historical cohort, 1960-1999," *Radiation Research* October 2005, Vol. 164 (4Pt1).

Plokhy, Serhii, *Chernobyl The History of a Nuclear Catastrophe,* New York: Basic Books, 2018.

Pollack, Jonathon D., *North Korea, Nuclear Weapons and International Security,* Oxon: The International Institute for Strategic Studies, 2011.

Potter, William C., "Potatoes Were Guarded Better," *Bulletin of the Atomic Scientists,* Vol. 51(3) May/June 1995.

Pourcelot, L., "Transfer factors of cesium in milk and meat determined from measurements made in France in 2011 following the Fukushim Daichi accident," *Radioprotection* Vol. 48 (3) April 2013.
Powaski, Ronald E. *Return to Armageddon: The United States and the Nuclear Arms Race, 1981-1999.* New York: Oxford University Press, 2000.
Qaiser, Rana, "Musharaf Pardons Quadeer on Cabinet Nod," *Daily Times*, Feb. 6, 2004.
Quinn, D. M., "The Japanese Tsunami and Resulting Nuclear Emergency at the Fukushima Power Facility: Technical, Radiologic and Response Perspectives," *Journal of Nuclear Medicine*, 52, Sept. 2011.
Rajan, M.S., *India's Foreign Policy and Relations*, New Delhi: South Asian Private, 1985.
Raman, M. V. "India and the Policy of No First Use of Nuclear Weapons," *Journal for Peace and Nuclear Disarmament*, 1(1), 2018
Rauf, Tariq, "The 2022 NPT Review Conference. From the Sublime to the Ridiculous," *Inter Press New Agency*, Friday, March 20, 2020.
Reed, Thomas C., *The Nuclear Express A Political History of the Bomb and its Proliferation.* Minneapolis: Zenith Press, 2009.
———. "The Chinese Nuclear Tests 1964-96," *Physics Today*, Vol. 61 (9), Sept. 2009.
———. *At the Abyss. An Insider's History of the Cold War.* New York: Ballantine Books, 2004.
Rhodes, Richard, *The Making of the Atom Bomb.* New York: Simon & Schuster, 1988.
———. "Why Did We Make Atomic Bombs?" James Conca Forbes, https: www.Forbes.com/sites/jamesconca.
———. *Dark Sun. The Making of the Hydrogen Bomb*, New York: Simon & Schuster, 1995.
———. *The Twilight of the Bombs. Recent Challenges, New Dangers and the Prospect of a World without Nuclear Weapons.* New York : Alfred A. Knopf, 2010.
Richelson, Jeffrey T. *Spying on the Bomb. American Nuclear Intelligence from Nazi Germany to Iran and North Korea.* New York : Norton, 2006.
Rindzeicute. Eyle, *The Power of Systems: How the Policy Sciences Opened up the Cold War*, Ebook, 2016.
Robock, Alan, "Multi-decadal global cooing and unprecedented ozone loss following a regional nuclear conflict," *Earth's Future* 2(4) 2014.
———. "Nuclear winter revisited with a modern climate model and current nuclear arsenals. Still catastrophic consequences," *Journal of Geophysical Research: Atmospheres* Vol. 112, No. D 13, July 6, 2007.
———. "Rapid expansion of nuclear arsenals by Pakistan and India threatens regional and global catastrophe," *American Geophysical Union*, Fall Meeting, Washington, DC, Dec. 2018.
———. "Asia treads the nuclear path unaware that self-assured destruction would follow from nuclear war," *J. Asian Studies*, 76, 2017.
Rosenau, William, Project Air Force (US): Rand Corporation, *Aum Shinrikyou, Al Qaeda, and the Kinshasa Reactor: Implications of three case studies for combating nuclear terrorism*, 2005.
Ross, Frank A., "Order from Chaos. Experts assess the nuclear Non-Proliferation Treaty, 50 years after it went into effect," *Brookings*, March 3, 2022.
Roussel-Debet, S. "Transfer of factors of cesium in milk and meat determined from measurements made in France following the Fukushima Daichi accident," *Radioprotection*, Vol. 48(3), April 2013.
Ruff, Tilman, "The Growing Threat of nuclear war and the role of the health community," *World Medical Journal* 62(3), April 2013.
Saecy, L., "Transfer factors of cesium in milk and meat determined from measurements made in France in 2011 following the Fukushima Daichi accident," *Radioprotection* Vol. 48(3), April 2013.
Scott-Clark, Catherine, *Deception Pakistan, the United States and the Secret Trade in Nuclear Weapons*, New York : Walker & Company, 2007.
Seed, David, "Suitcase Nukes," In: *US Narratives of Nuclear Terrorism Since 9/11 Worst Case Scenarios*, Liverpool, UK: Palgrave, 2003.

Segerstahl, Boris, "The Long Shadow of Soviet Plutonium Production," *Environment* Vol. 39, No. 1, January 1, 1997.

Shaikh, Farzana, "Pakistan's nuclear bomb: beyond the non-proliferation regime," *International Affairs* (London), January 2002 Vol. 78 (1).

Shaker, Mohammed I., *The Nuclear Non-Proliferation Treaty. Origins and Implementation 1959-1979,* 3 Volumes, London: Oceana Publications, 1980.

Shanty, Frank, "Has Al Qaeda Acquire Nuclear Weapons? The State of Existing Knowledge," In: Shanty, Frank and Misra, Patit Paban, *Organized Crime from Trafficking to Terrorism,* ABC-CLIO, LC, 2007.

Shellenberger, Michael, "If nuclear plants are so vulnerable to terrorist attack, why don't terrorists attack them?" *Forbes,* July 6, 2018.

Sherwin, Martin, *A World Destroyed. The Atom Bomb and the Grand Alliance.* New York: Knopf, 1975.

Shieber, Jonathan, "North Korean Hackers Stole South Korean and US War Plans," *TechCrunch,* Oct. 10, 2017.

Shields, John M. (ed.), *Dismantling of Cold War US and NIS Perspectives on the Nunn-Lugar Cooperative Threat Reduction Program.* Cambridge, MA: MIT Press, 1997.

Shunya, Yoshimi, "Radioactive Rain and the American Umbrella," *Journal of Asian Studies,* Vol. 71 (No. 2), May 2012.

Sich, Alexander R., "Truth was an Early Casualty," *Bulletin of the Atomic Scientists,* Vol. 52, No. 3, May 1, 1996.

Siddique, F., "Pakistan's Strategic Nuclear Policy and Implications for Deterrance Stability," *CISS Insight : Quarterly News and Views,* IV(1), 2016.

Simon, Steven L., "Radiation doses to local population near nuclear weapons test sites worldwide," *Health Physics* 82 (5), 2002.

———. "Health Effects of nuclear weapons testing," *The Lancet,* 1-7 Aug. 2015, Vol. 386 (9992).

———. "Fallout from Nuclear Weapons Tests and Cancer Risks," *American Scientist* Vol. 94, No. 1, Jan. 2006.

Smigielski, David, "A Review of the Suitcas Nuclear Bomb Controversy," *Partnership for Global Security,* September, 2003.

Smith, Jacqueline, "North Korea: The Next Nuclear Nightmare?" *Arms Control Today,* Vol. 21, Number 2 (March 1991).

Smith-Norris, Martha, "Only as Dust in the Face of the Wind. An Analysis of the BRAVO Nuclear Incident in the Pacific, 1954," *Journal of American East Asian Relations,* Vol. 6, No. 1 (Spring 1997).

Smith, R. Jeffrey, "Pakistan nuclear scientists' accounts tell of Chinese proliferation," www//washingtonpost.com/wp-dyn/content/article/2009/11/12/AR2009-111211060.html.

Sokoloski, Henry D., *Best of Intentions America's Campaign against strategic weapons.* Westport, Connecticut: Praeger, 2001.

———. "Serious Rules for Nuclear Power without Proliferation," *Nonproliferation Review* Vol. 40 (25), July 2015.

Soto-Iglesias, M., "An analysis of the hydrogen explosion in the Fukushima Daichi accident," *International Journal of Hydrogen Energy,* Vol. 40(25), July 2015.

Spector, Leonard, "North Korea: The Next Nuclear Nightmare?" *Arms Control Today,* Vol. 21, Number 2 (March 1991).

Steele, Karen Dorn, "Hanford's Bitter Legacy," *Bulletin of the Atomic Scientists,* Vol. 44, No. 1, January-February 1988.

Stenchikov, G. I., "Atmospheric effects and societal consequences of regional nuclear conflicts and acts of individual nuclear terrorism," *Atmos. Chem. Phys.,* 7, 1973-2002, 2007.

_____"Nuclear winter revisited with a modern climate model: Still catastrophic consequences," *Journal of Geophysical Research. Atmospheres* Vol. 112, No. D. 13, July 6, 2007.

Stepanenko, Valery, "The Influence of the Lop Nor Nuclear weapons test base on the population of the Republic of Kazakhstan," *Radiation Measurements,* Vol. 46(4), 2011.

Stetcu, Ionel, "Nuclear Fission Dynamics: Past, Present Needs, Future," *Frontiers in Physics.* doi: 10: 3381/phy 2020.0063

Stillman. Danny B. *The Nuclear Express: A Political History of the Bomb and Its Proliferation.* Minneapolis: Zenith Press, 2009.

Strauss, H. W., "The Japanese Tsunami and Resulting Nuclear Emergency at the Fukushima Power Facility: Technical, Radiological and Response Perspectives," *Journal of Nuclear Medicine,* v. 52, Sept. 2011.

Sugdeev, Roald, "Reconstructed Account of the conversation between Niels Bohr and Y.P. Terletsky," In: Rhodes, *Dark Sun,* 1995.

Sundaram, K., "India and the Policy of No First Use of Nuclear Weapons," *Journal for Pearce and Nuclear Disarmament,* Vol. 1(1), 2018.

Sung-hun, Choe, "North Korea Fires Missile over Japan," *New York Times,* Aug. 28 2014.

———. "North Korean Defectors Show Signs of Possible Radiation Exposure," *New York Times,* Dec. 27, 2017.

Szilard, Leo, "My Trial as a War Criminal," In: *The Voice of the Dolphin and Other Stories,* New York: Simon & Schuster, 1961.

Taira, W., "The Fukushima nuclear accident and the pale grass blue butterfly; evaluating biological effects of long-term low dose exposures," *BMC Evolutionary Biology,* 13: 168, 201: C3.

Tanaka, Miyazawa, "The Influence of the Lop Nor Nuclear Weapons Test Base on the population of the Republic of Kazakhstan," *Radiation Measurements* Vol. 46 (4), 2011.

Tannenwald, Nina, "The Vanishing Nuclear Taboo? How Disarmament Fell Apart," *Foreign Affairs,* V. 97, Number 6, November/December 2018.

———. *The Nuclear Taboo. The United States and the Non-Use of Nuclear Weapons since 1945.* New York: Cambridge University Press, 2007.

Taylor, Julia-Lee, "Multidecadal global cooling and unprecedented ozone loss following a regional nuclear conflict," *Earth's Futures* 100, Vol. 2, Issue 4, April 2014.

Tenet, George J., Director of Central Intelligence Agency, testimony before the Senate Armed Services Committee, March 19, 2002.

Thomas, Raju G. C. (ed.) *India's Nuclear Security.* London: Lynne Rienner Publishers, 2000.

Thomas, Steve, "What will the Fukushima disaster change? (Viewpoint), *Fukushima Daichi nuclear accident in Japan Report, Energy Policy,* Vol. 45, June 2012.

Titiuji, Chouchou, "The Influence of the Lop Nor Nuclear Weapons Test Base on the population of the Republic of Kazakhstan," *Radiation Measurements* Vol. 46(4), 2011.

Tkachenko, Yevgeny, "FSB Prevent Theft of Nuclear Materials," ITAB-TASS, Dec. 8, 1998.

"Minatom Says 1998 Theft Involved HEU," *Centre for Nonproliferation Studies, NIS,* Trafficking Data Base, Oct. 31, 2000.

Tomiya, Atsushi, "Radiological impact of the nuclear power plant disaster on freshwater fish. An overview of monitoring results," *Journal of Environmental Radioactivity,* Vol. 151, Part 1, Jan. 2016.

Toon, Owen B., "Multideadal global cooling and unprecedented ozone loss following a regional nuclear conflict," *Earth's Future,* 100, Vol. 2, Issue 4, April 2014.

"Atmospheric effects and societal consequences of regional-scale nuclear conflicts and acts of terrorism," *Atmos. Chem. Phys.* 7, 1973-2002, 2007.

"Asia treads the nuclear path, unaware that self assured destruction would result from nuclear war," *J. Asian Studies,* 76, 2017.

"Rapid expansion of nuclear arsenals by Pakistan and India threatens regional and global catastrophe," *American Geophysical Union,* Fall meeting, Washington DC, Dec. 2018.

Toyoda, Shin Endo, "The Influence of the Lop Nor Nuclear Weapons Test Base on the population of the Republic of Kazakhstan," *Radiation Measurements*, Vol. 46 (4), 2011.

Tsumune, Daisuke, "Distribution of 137 CS and 134 S in the North Pacific Ocean: impacts of the TEPCO Fukushima Daichi NPP accident," *Journal of Radioanalytical and Nuclear Chemistry*, Vol. 296 (1), 2013.

Turco, R. P., "Atmospheric effects and societal consequences of regional-scale nuclear conflicts and acts of individual nuclear terrorism," *Atmos. Chem. Phys.* 7, 1973-2002, 2007.

Tuttle, R. M., "The Japanese Tsunami and Resulting Nuclear Emergency at the Fukushima Daichi Power Facility: Technical, Radiologic and Respons Perspectives," *Journal of Nuclear Medicine*, 52, Sept. 2011.

Ur-Rehman. Sahid, *The Long Road to Chagai*, Islamabad: Printwise Publications, 1999.

Vivek, Raghuvanshi, "India to Stay the Course on Nuke Doctrine," *Defence News*, 1 November 2004. :

Von Hippel, Frank N., "The Decision to End US Nuclear Testing," *Arms Control Today* Dec. 2019.

Von Neumann, Thomas E., "Radioactive Debris from Operation Castle. Worldwide Fallout," New York: United States Atomic Energy Commission, OSTI.GOV, 1955.

Wada, Toshihiro, "Radiological impact of the nuclear power plant accident on freshwater fish: An overview," *Journal of Environmental Radioactivity* Vol. 151, Part 1, Jan. 2016.

Warrick, Joby, "Pakistan's nuclear scientists' accounts tell of Chinese proliferation," *Washington Post*, Dec. 11, 2009.

Watson, Peter, *Fallout. Conspiracy, Cover-up, and the Deceitful Case for the Atom Bomb. A New History*. New York: Public Affairs, 2019.

Willrich, Mason, "No First Use of Nuclear Weapons: An Assessment," *Orbis* Vol. 9, No. 2 (Summer 1965).

Wilson, Richard, "A Visit to Chernobyl," *Science* 236 No. 4809, 1987.

Winterer, Edward J. "Nuclear Bomb Testing," In: *Encyclopedia of Islands*, Ed. Gillespie, Rosemary G. and Claque, David A., Berkeley CA: University of California Press, 2009.

Wohlstetter, Albert, "The Military Potential of Civilian Nuclear Energy," *Minerva*, Vol. 15, Issue 3 (Autumn-Winter 1977).

Xia, Lili, "Asia treads the nuclear path, unaware that self-assured destruction would result from nuclear war," *J. Asian Studies* 76, 2017.

"Rapid expansion of nuclear arsenals by Pakistan and India threatens regional and global catastrophe," *American Geophysical Union*. Fall Meeting, Washington DC, Dec. 2018 GC33B-12.

Yablokov, Alexy V. *Chernobl Consequences of the Catastrophe for People and the Environment*, Annals of the New York Academy of Sciences, Vol. 1181, Boston: Blackwell, 2009.

Yamazaki, Hideo, "Radioactive contamination in the Tokyo metropolitan area in the early stage of the Fukushima Daiichi Nuclear Power Plant (FDNPP) accident and its fluctuation over five years," PLosONE, Vol. 12 (11), p.e. 018 7687, Nov. 14, 2017.

Yonez, J., "An analysis of the hydrogen explosion in the Fukushima Daichi accident," *International Journal of Hydrogen Energy*, 6 July 2015, Vol. 40 (25).

Yoshida. Sumito, "The Communication of Information Such as Evacuation Orders at the time of a Nuclear Power Station Accident, *Japan Medical Association Journal*, 57 (5-6), Dec. 2017.

Zanzonico, P., "The Japanese Tsunami and Resulting Nuclear Emergency at the Fukushima Daiichi Power Facility: Technical, Radiologic and Response Perspectives," *Journal of Nuclear Medicine* 52, Sept. 2011.

Zharlyganova, Dinara, "The Influence of the Lop Nor Nuclear Weapons Test Base on the population of the Republic of Kazakhstan," *Radiation Measurements* Vol. 46(4), 2011.

Zhaxybay, Apsalikov, "The Influence of the Lop Nor Nuclear Weapons Test Base on the population of the Republic of Kazakhstan," *Radiation Measurements* Vol. 46(4), 2011.

Zhumadilov, Kassyn, "The Influence of the Lop Nor Nuclear Weapons Test Base on the population of the Republic of Kazakhstan," *Radiation Measurements* Vol. 46(4), 2011.

Index

About the Author

John Perry is a Jesuit priest and academic who has previously published books on challenging realities in human life: torture, the slave trade, and epidemics in countries unable to deal with them. He tries to help his readers understand these phenomena, what leads to them, and what we can do about them. His works include *Torture: Religious Ethics and National Security* (2005), *Catholics and Slavery. A Compromising History* (2008), and *Social Mobilization and the Ebola Virus* (2016, Lexington).

www.ingramcontent.com/pod-product-compliance
Ingram Content Group UK Ltd.
Pitfield, Milton Keynes, MK11 3LW, UK
UKHW041843031026
15587UKWH00002B/972